AF886316

Science Pop Art
Astronomy & Space

A Terminology Coloring Book

Illustrated and Written By

Mizzz Foster

Published by Fantastic Absurdities
Lake Elsinore, CA US
Email: MizzzFoster@gmail.com

ISBN:
Printed in the United States of America
Design by M. A. Foster
Cover Design by M. A. Foster

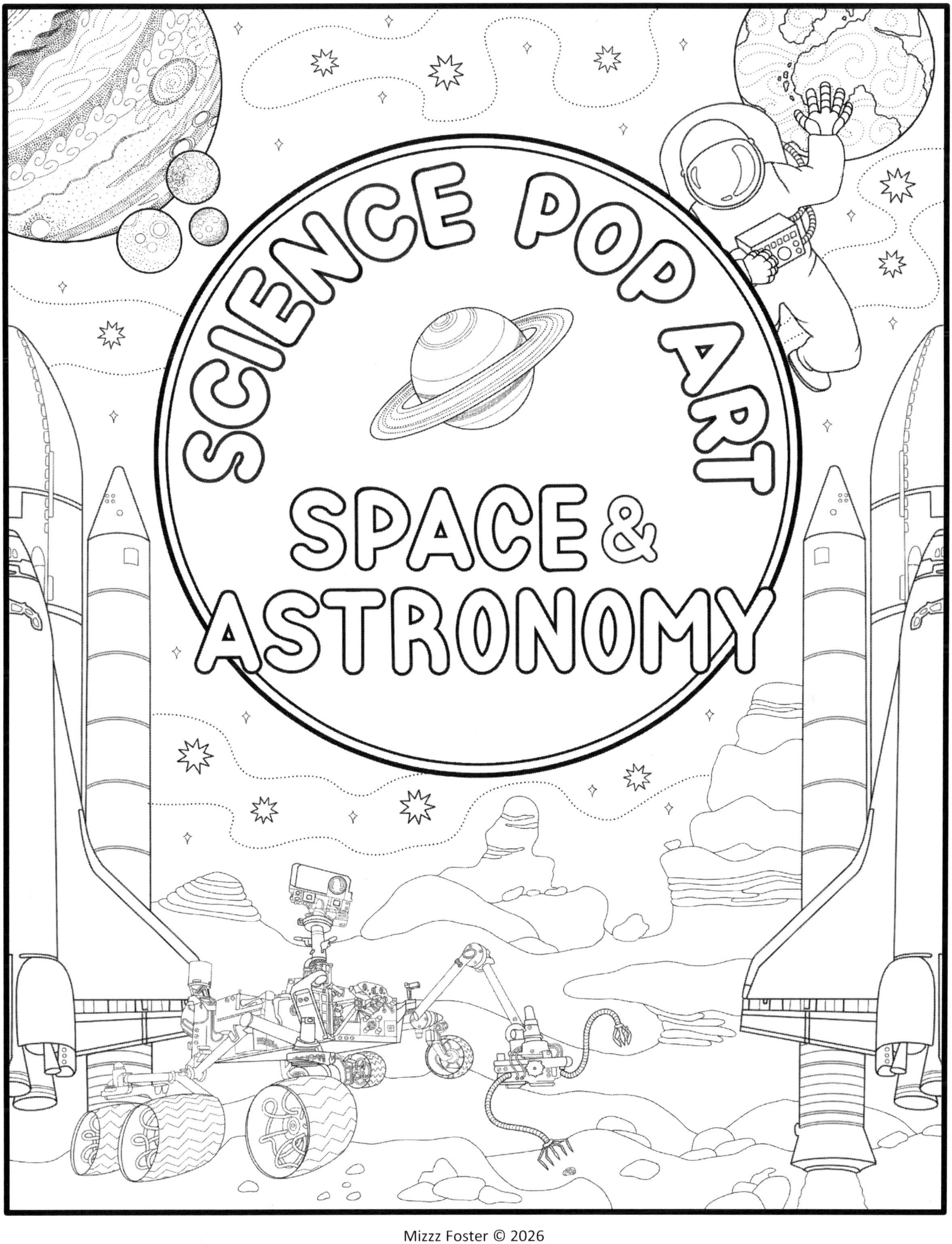
SCIENCE POP ART
SPACE & ASTRONOMY

ASTRONOMY

Astronomy is the scientific study of the Sun, Moon, stars, planets, and other celestial bodies in space. Unlike astrology, which is based on beliefs, astronomy is a science that uses observation, experimentation, and mathematics to understand the universe.
While astronomy and astrology both involve the study of celestial objects, they are very different. Astronomy is a science that explains how stars, planets, and galaxies behave using evidence, experimentation, and exploration. Astrology is a belief system that claims the positions of stars and planets on a person’s birthday, can influence people's personalities and predict events on Earth, but it is not based on scientific evidence. Astronomy seeks to understand the universe through careful observation, while astrology is rooted in tradition and interpretation.

For thousands of years, human civilizations around the world have studied the sky, adding and layering the foundations for the science of astronomy that we know today. Early civilizations developed detailed sky-watching systems long before telescopes existed. Ancient people used the stars to track time, predict seasons, guide navigation, and plan religious festivals.

African Contributions to Early Astronomy

In northeastern Africa, the Nabta Playa site, located in the Nubian Desert near the Egypt-Sudan border, is one of the oldest known astronomical observatories, dating back to around 5000 BCE. The stone circle and alignments found there may have helped ancient peoples track the summer solstice and the seasonal rains, crucial for survival in the desert.

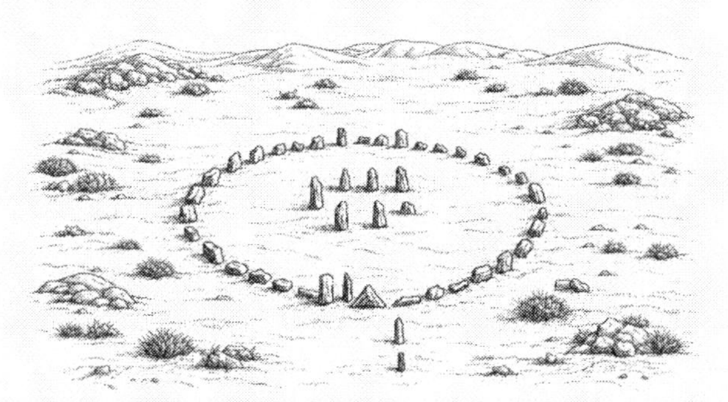

NABTA PLAYA

In Ancient Egypt, astronomy played a vital role in agriculture, religion, and architecture. The Egyptians observed the night sky to develop one of the world’s first 365-day calendars, which was closely tied to the heliacal rising of Sirius, a bright star that signaled the annual flooding of the Nile River. Temples and pyramids were often aligned with celestial bodies, demonstrating the Egyptians’ deep connection between the heavens and the Earth.

Farther west, the Dogon people of Mali developed sophisticated star knowledge, particularly about Sirius, the brightest star in the sky. Their oral traditions describe not only Sirius A but also an invisible companion star, Sirius B, information that Western astronomers didn’t confirm until the 1800s. Whether through careful observation or cultural exchange, the Dogon's astronomical knowledge reflects generations of sky study and cultural memory.

Ancient Chinese Astronomy

In Ancient China, astronomy was highly advanced and closely linked to the emperor, agriculture, and society. Chinese astronomers recorded eclipses, comets, sunspots, and novae for thousands of years. They divided the sky into detailed star charts and developed an early system of celestial coordinates. The Chinese used instruments like armillary spheres and water clocks to study the movement of stars and planets with incredible accuracy.

An armillary sphere is an ancient astronomical instrument made of rings and circles representing important parts of the sky, like the celestial equator, ecliptic, and meridians. A water clock is one of the oldest time-measuring instruments, using the steady flow of water to keep track of time. Observatories were often state-funded, and astronomical knowledge was essential to maintaining the emperor's authority, as celestial signs were seen as reflections of the kingdom's fortune and the emperor's destiny.

Observatories of the Americas and Europe

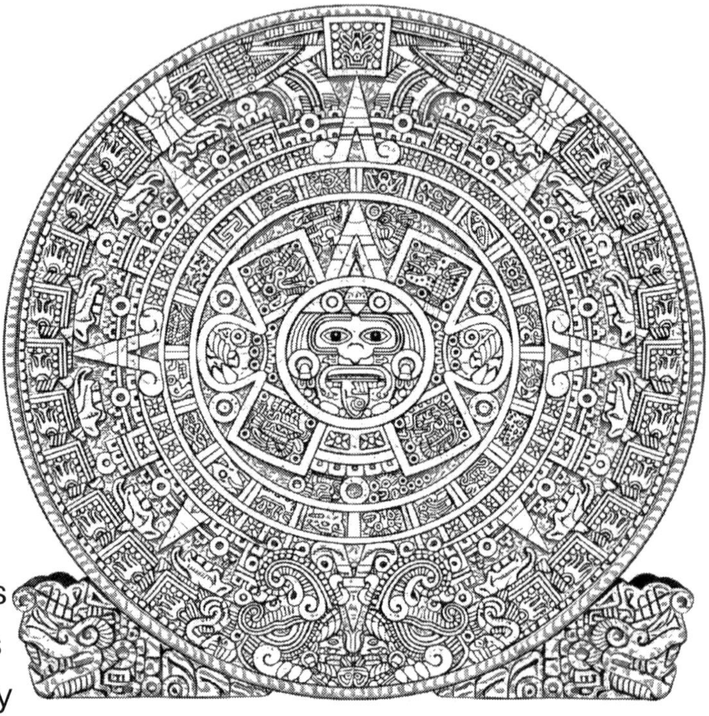

In the Americas, the Maya civilization of Central America tracked celestial movements with extraordinary precision. They built observatories like El Caracol in Chichén Itzá, developed complex calendar systems, and closely followed the motion of Venus, the Moon, and eclipses to guide religion, agriculture, and warfare.

In prehistoric Britain, Stonehenge stands as a mysterious monument possibly used to track the summer and winter solstices. The massive stones align with the sunrise and sunset on specific days of the year, showing that early people used the sky to mark seasonal changes and timekeeping.

Famous Astronomers and the Evolution of Astronomy

Many famous astronomers have shaped what we know today. In ancient Greece, Ptolemy proposed the geocentric model, where Earth was the center of the universe. During the Renaissance, Nicolaus Copernicus introduced the heliocentric model, placing the Sun at the center of the solar system. Galileo Galilei improved the telescope and discovered moons orbiting Jupiter, proving not everything orbits Earth. Johannes Kepler explained that planets move in elliptical orbits, and later, Edwin Hubble discovered that the universe is expanding, which led to the Big Bang theory.

Modern Astronomy

Astronomy has changed greatly over time. In the past, people observed the sky with the naked eye while developing monuments and rudimentary equipment. Today, scientists use space telescopes, satellites, and computer modeling to study the universe. We now explore distant galaxies, discover black holes, identify exoplanets, and even consider the possibility of life beyond Earth. Astronomy has grown from ancient stargazing into a high-tech science that helps us understand our place in the cosmos.

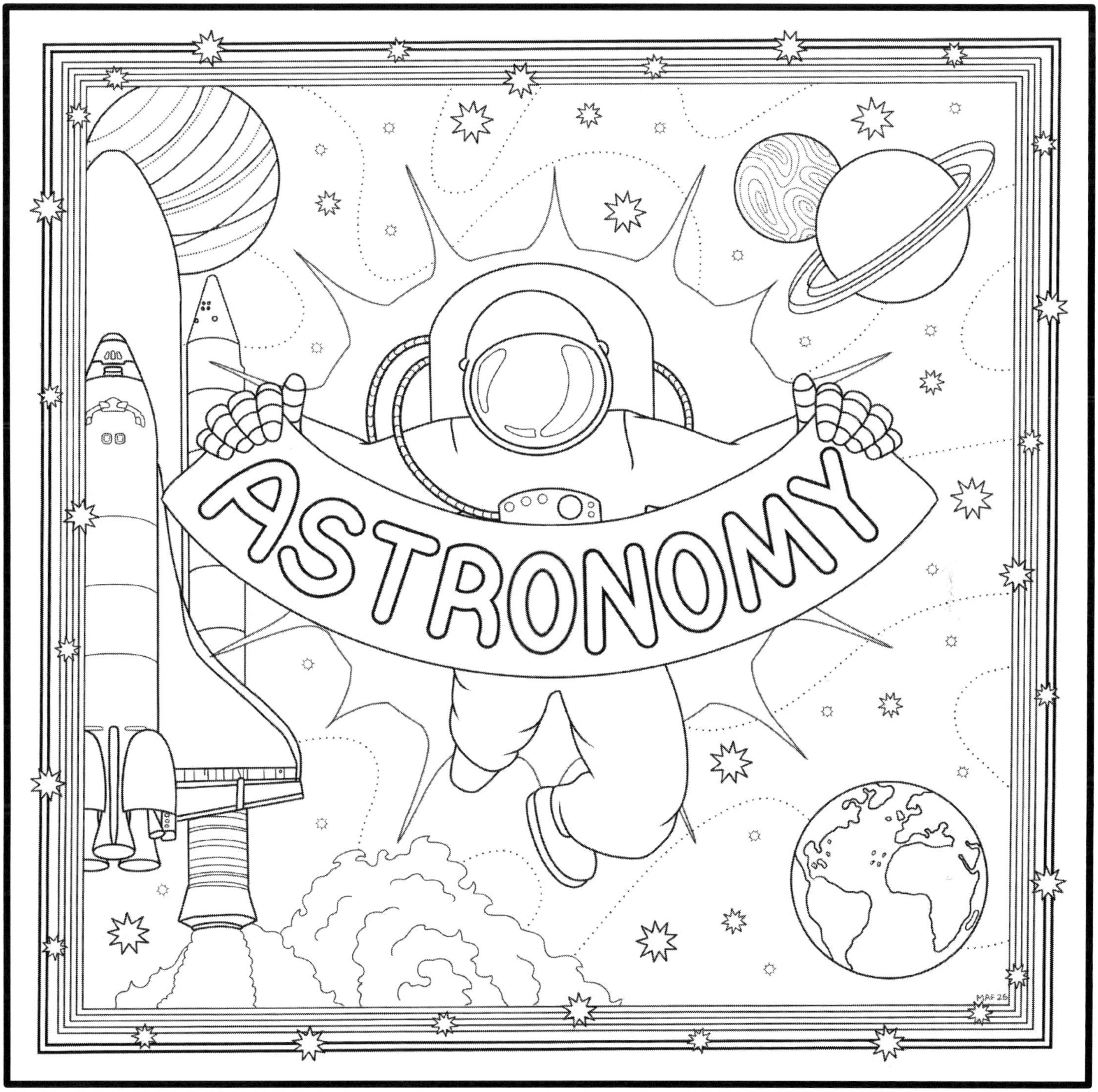

What are the historical origins of astronomy?	What is astronomy, and how is it different from astrology?
Who two famous astronomers and what are their contributions?	What tools and technologies do astronomers use to study space?

SPACE

Space, also called outer space, is the vast region beyond Earth's atmosphere where planets, stars, galaxies, and other celestial objects exist. It begins at the edge of Earth's atmosphere and extends throughout the entire universe. Space is not completely empty, but it contains extremely small amounts of matter, including gas, dust, radiation, and energy. Because space has no breathable air and very little pressure, it is considered a near vacuum.

In space, gravity plays a major role in how objects move and interact. Gravity keeps planets in orbit around stars and moons in orbit around planets. Large collections of stars, gases, and dust are held together by gravitational fields of larger celestial bodies to form galaxies. Although gravity is weaker in space than on Earth, it still controls the motion of objects over vast distances and shapes the structure of the universe.

Space is the setting for many important astronomical processes. Stars form in large clouds of gas and dust called nebulae. Inside stars, nuclear fusion produces energy that creates light and heat. Over time, stars change as they use up their fuel, eventually becoming objects such as white dwarfs, neutron stars, or black holes. These processes occur entirely in space and help scientists understand how matter and energy behave under extreme conditions.

Astronomers study space using telescopes and spacecraft that observe objects across many wavelengths of electromagnetic radiation, including visible light, infrared, X-rays, and radio waves. Studying space allows scientists to learn about the origin of the universe, the life cycles of stars, and the possibility of life beyond Earth. Observations of space also help improve technology, communication systems, and our understanding of Earth's place in the universe.

Space is important to astronomy because it is where all cosmic events take place. From the motion of planets to the formation of galaxies, space provides the environment in which the universe exists and evolves. By studying space, scientists continue to expand knowledge about how the universe began, how it changes, and how it may develop in the future.

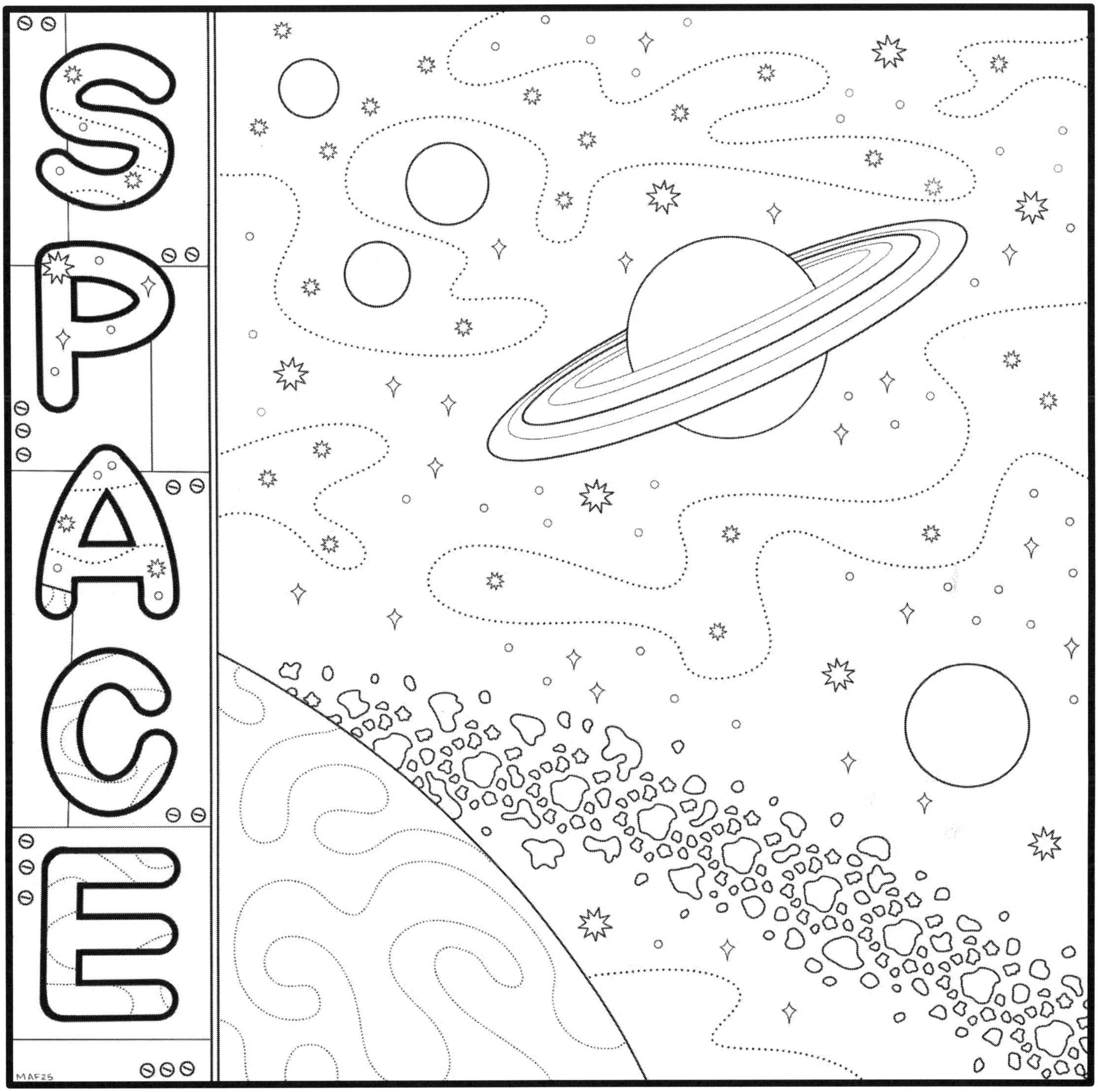

What is outer space, and why is it considered a near vacuum?	How does gravity influence objects and structures in space?
Name one important process that occurs in space and explain its significance.	Why is studying space important for understanding the universe and Earth's place in it?

GALAXY

A galaxy is a massive system made up of billions or even trillions of stars, along with planets, gas, dust, and dark matter, all held together by gravity. Galaxies are the basic building blocks of the universe. They vary greatly in size and shape, ranging from small dwarf galaxies to enormous galaxies that span hundreds of thousands of light-years.

Galaxies form when gravity pulls large clouds of gas and dust together over long periods of time. As these clouds collapse, stars begin to form, and their gravity attracts more matter. Over billions of years, this process creates structured systems of stars, planets, and stellar remnants that orbit a common center. Most large galaxies have a supermassive black hole at their center, which influences the motion of stars nearby.

Astronomers classify galaxies into several main types based on their shape and structure. Spiral galaxies have flat, rotating disks with curved arms extending outward from a central bulge. Elliptical galaxies are more rounded or oval-shaped and contain older stars with little gas or dust for new star formation. Irregular galaxies do not have a defined shape and are often the result of gravitational interactions or collisions with other galaxies.

Galaxies are where stars are born, evolve, and die. Many elements essential for planets and life are created inside stars and distributed throughout galaxies when stars explode or release material. By studying galaxies, astronomers learn about the history of the universe, how matter is distributed across space, and how cosmic structures change over time.

Galaxies are not isolated objects. They often exist in groups or clusters and can interact with one another through gravity. These interactions can trigger bursts of star formation, distort galactic shapes, or even cause galaxies to merge. Studying these interactions helps scientists understand how galaxies grow and evolve across billions of years.

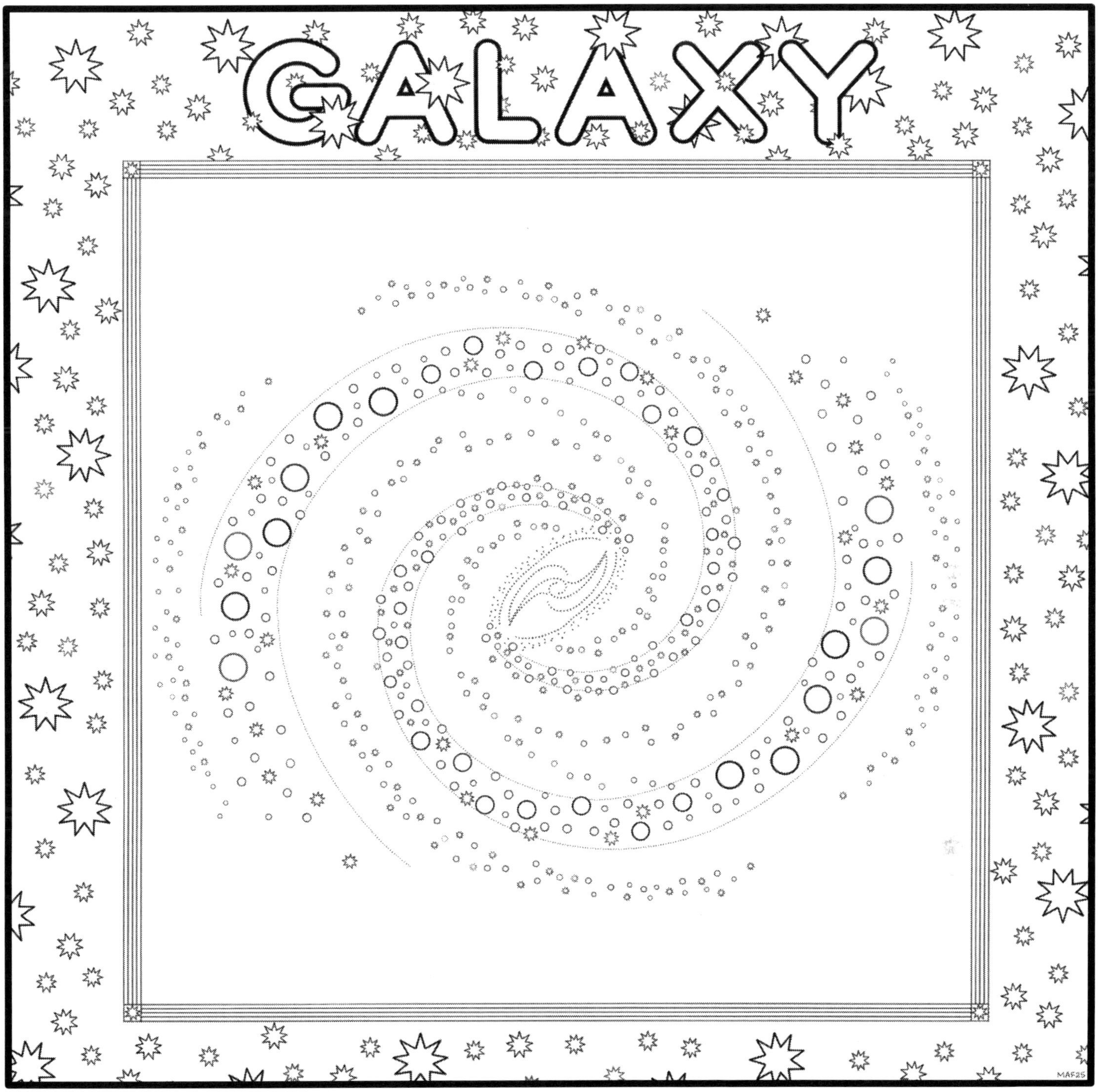

What is a galaxy, and what types of objects does it contain?	How do galaxies form and remain held together?
Describe one type of galaxy and its characteristics.	Why is the study of galaxies important to understanding the universe?

SOLAR SYSTEM

The solar system formed about 4.6 billion years ago from a massive cloud of gas and dust called a solar nebula. A nebula is a vast, interstellar cloud of dust, hydrogen, helium, and other ionized gases. Gravity pulled most of this material toward the center, forming a magnificent star, which we call the Sun. The remaining matter flattened into a spinning disk. Over time, tiny particles collided and stuck together, gradually building the planets, moons, asteroids, and comets. This slow process of growth and change shaped the orderly yet complex and diverse system that surrounds our Sun.

Our solar system is made up of the Sun, eight planets, their moons, dwarf planets, asteroids, comets, and countless smaller pieces of rock, ice, and dust. The inner planets—Mercury, Venus, Earth, and Mars—are rocky worlds, separated from the outer planets by the asteroid belt. The outer planets include massive gas giants and icy giants with complex ring systems and dozens of moons. Beyond the planets lie regions such as the Kuiper Belt, and the distant Oort Cloud, which hold remnants from the solar system's earliest days along with slow orbiting comets.

Scientists study the solar system using a wide variety of tools and instruments. Telescopes on Earth and in space allow astronomers to observe planets, moons, and distant objects in visible light and other wavelengths. Space probes, orbiters, landers, and rovers travel directly to other worlds, collecting images, chemical data, and physical measurements. Instruments such as spectrometers, cameras, radar, and particle detectors help scientists analyze surfaces, atmospheres, magnetic fields, and motion.

The history of the solar system is constantly evolving because new discoveries are made every year. Improved technology, powerful space missions, and fresh data continue to reshape what scientists understand about planet formation, planetary motion, and the potential for life beyond Earth. Objects once thought to be simple are now known to be complex and active, and new worlds are still being discovered and forming. The solar system is not a finished story, it is an ongoing exploration, full of mystery, motion, beauty, and wonder.

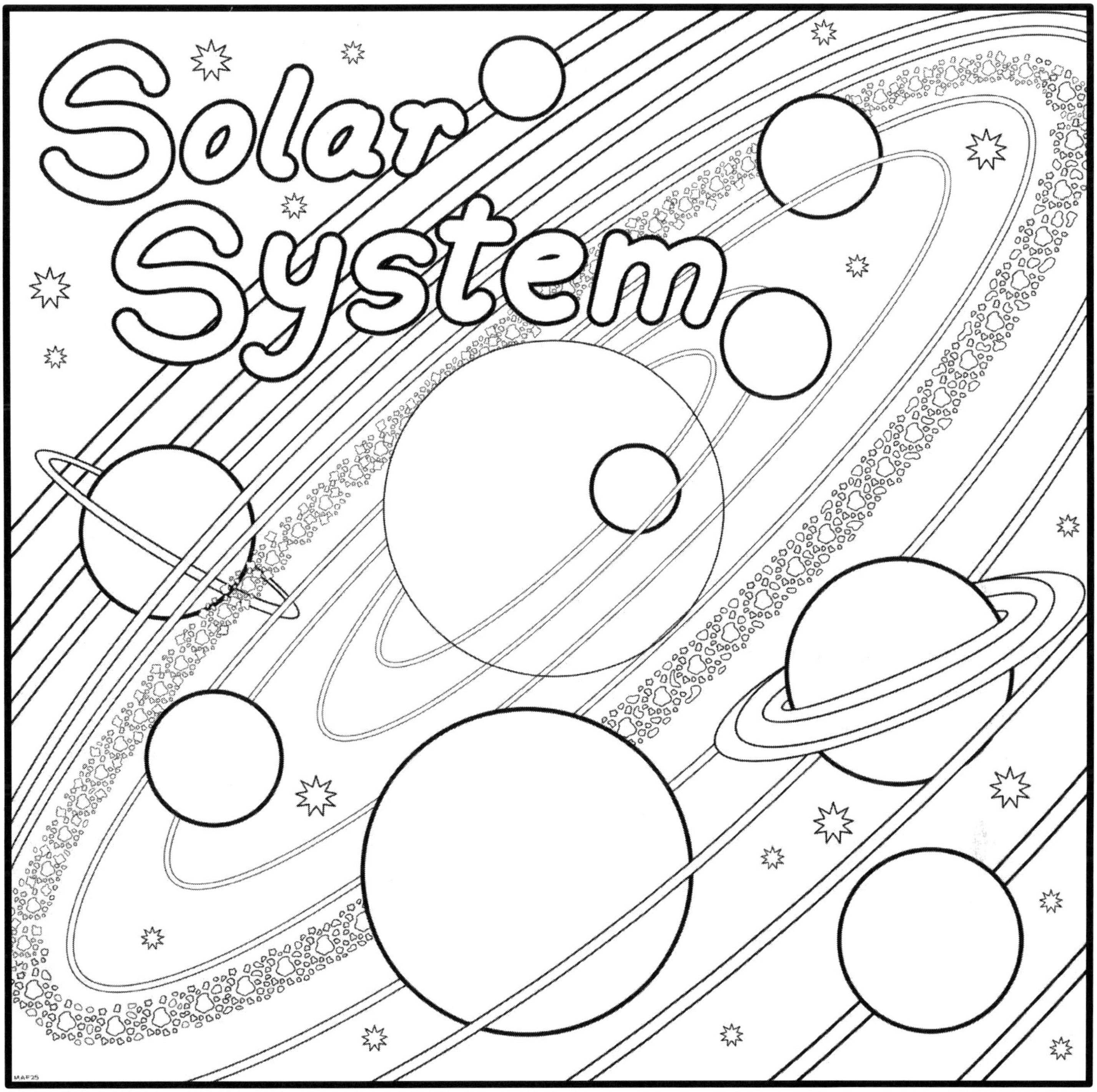

What are the main components of our solar system?	How do the inner planets differ from the outer planets?
What separates the inner planets from the outer planets?	How would the sun be classified?

THE SUN

History and Discovery

The Sun has been observed by humans since the beginning of history because it is the most important and visible object in the sky. Ancient civilizations worshiped it as a powerful deity, and many early cultures, built calendars and monuments to track its motion. Scientists such as Copernicus and Galileo helped show that Earth and the other planets revolve around the Sun.

Size

The Sun is by far the largest object in the solar system. Its diameter is about 1.39 million kilometers (865,000 miles), which is roughly 109 times the diameter of Earth. Over one million Earths could fit inside the Sun.

Gravity

The Sun's gravity is incredibly strong at about 274 m/s^2 on its surface, nearly 28 times stronger than Earth's gravity. Gravity is proportional to size. The larger a celestial body is, the larger its force of gravity will be. Its powerful gravitational pull holds all the planets, moons, asteroids, and comets in orbit within the solar system.

Composition

The Sun is a giant ball of hot gases, mostly hydrogen (about 74%) and helium (about 24%), with small amounts of heavier elements. Inside, nuclear fusion reactions convert hydrogen into helium, releasing enormous amounts of energy in the form of light and heat.

Location in the Solar System

The Sun is located at the center of the solar system, and all planets orbit around it.

Viewing the Sun in the Sky

The Sun can be viewed in the sky during the daytime. However, it is very dangerous to look directly at the Sun without proper protection because its light can damage the eyes. During rare solar eclipses, the Sun can be safely viewed using special equipment.

Name and Meaning

The word "Sun" comes from Old English and Germanic roots meaning "the bright one." In Roman mythology, the Sun was associated with the god Sol, and in Greek mythology, with the god Helios. Its name reflects its role as the source of light and life on Earth.

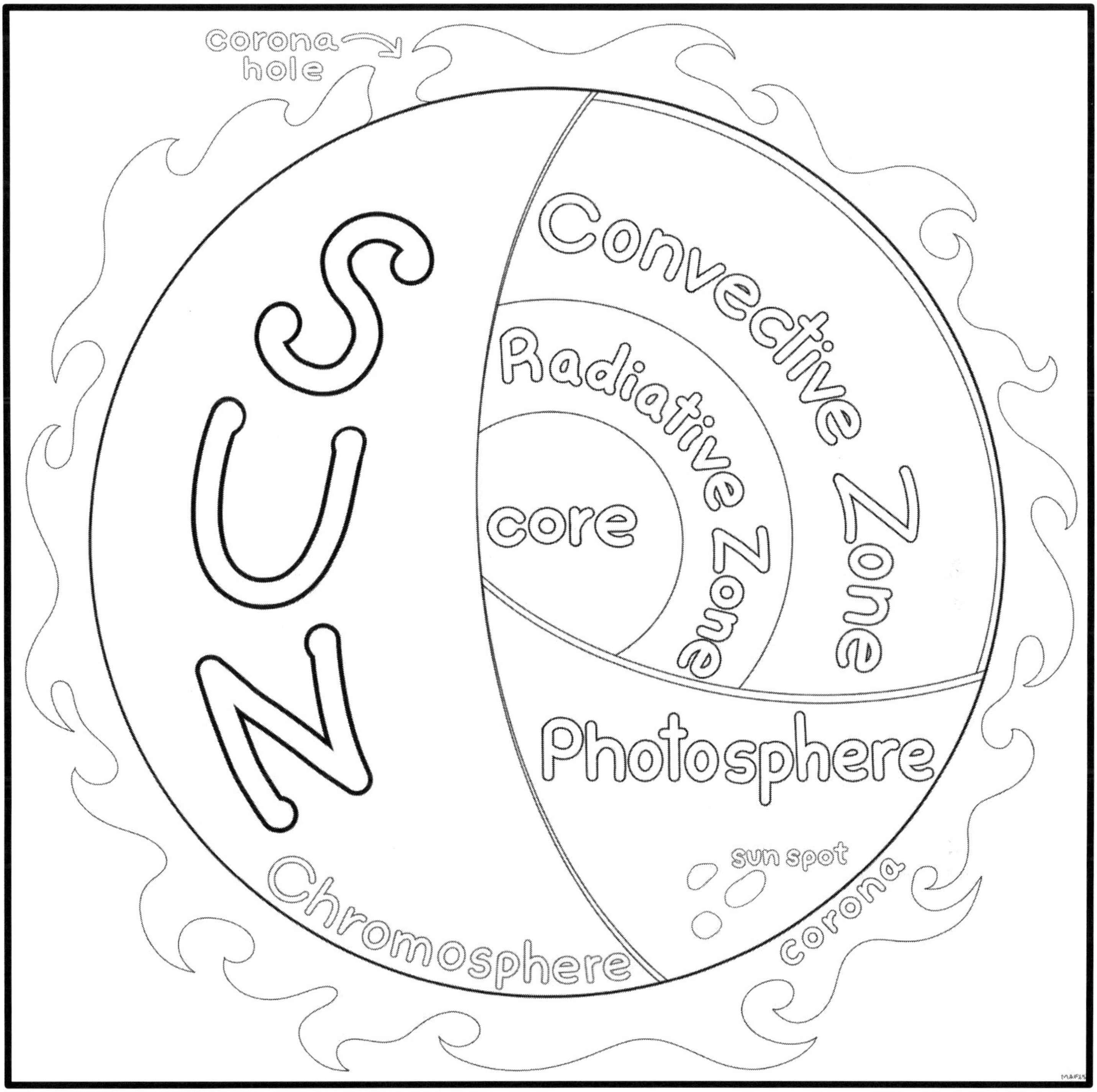

How did ancient civilizations and early scientists understand and study the Sun differently over time?	Why is the Sun's gravity important to the structure and stability of the solar system?
What is the Sun made of, and how does it produce the energy that reaches Earth?	How does the Sun's size and location influence its role in the solar system?

PLANET

A planet is a large celestial body that orbits a star and does not produce its own light. Planets are visible because they reflect light from the star they orbit. In our solar system, planets orbit the Sun and follow stable, predictable paths controlled by gravity. Planets are massive enough for gravity to pull them into a nearly spherical shape, which distinguishes them from smaller objects such as asteroids.

In astronomy, a planet is defined by several key characteristics. A planet must orbit a star, have enough mass for its gravity to shape it into a rounded form, and have cleared most other objects from its orbital path. Clearing the orbit means the planet has become gravitationally dominant in its region of space, either pulling in, pushing away, or capturing nearby objects. This distinguishes planets from dwarf planets, which share their orbits with other debris.

Planets vary greatly in size, composition, and structure. In our solar system, the inner planets are rocky and have solid surfaces, while the outer planets are gas or ice giants with thick atmospheres and no solid surface. Some planets have moons, rings, strong magnetic fields, or active weather systems. Despite these differences, all planets are shaped by the same physical forces, including gravity, motion, and energy from their parent star.

Planets are important to the study of astronomy because they provide insight into how solar systems form and evolve. By studying planets in our solar system and planets orbiting other stars, called exoplanets, scientists learn how common planetary systems may be throughout the universe and what conditions might support life. Understanding planets helps astronomers better understand Earth's place in space and the processes that shape worlds beyond our own.

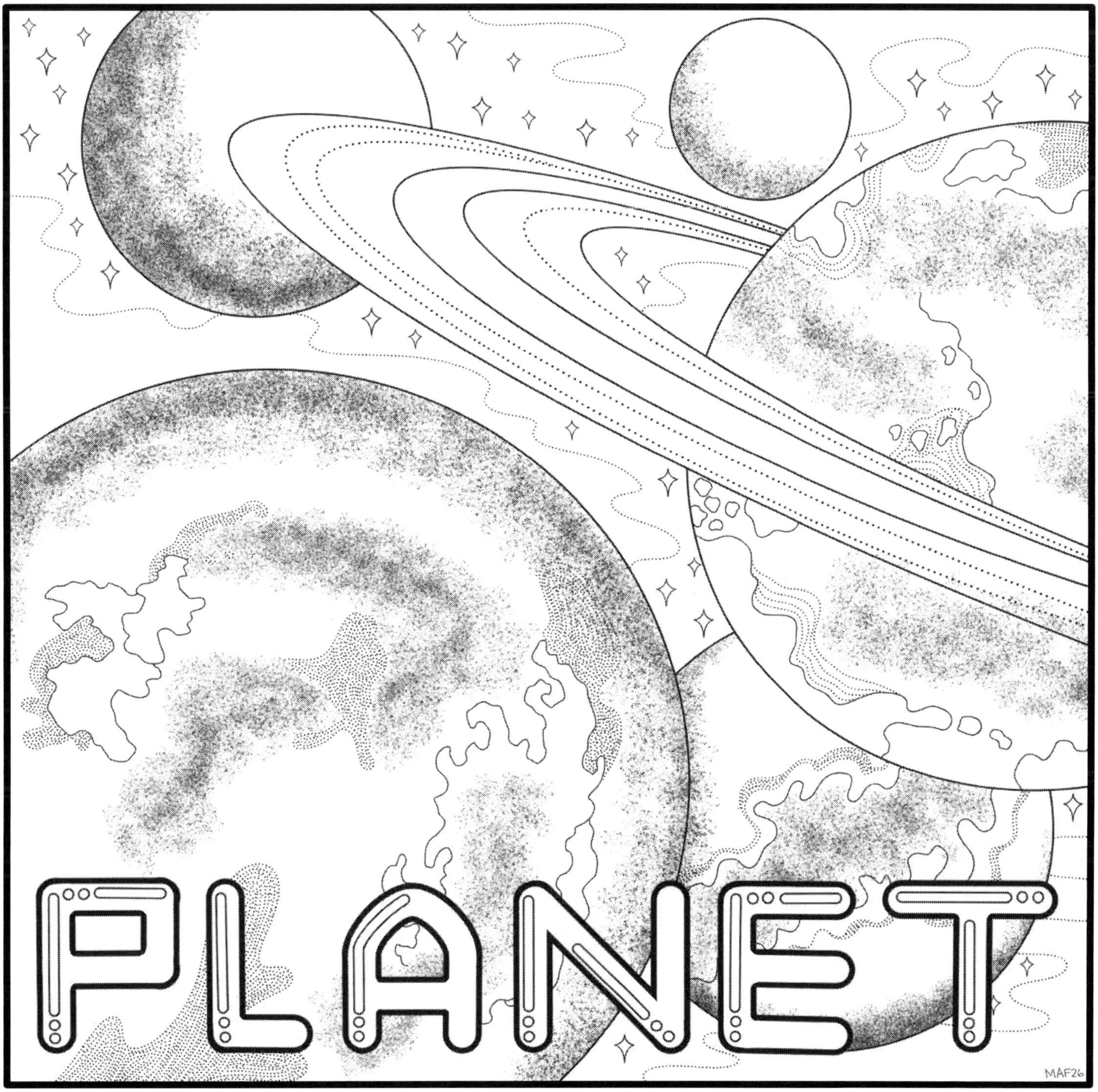

What is a planet, and why do planets not produce their own light?	What three main characteristics are used to define a planet in astronomy?
How do inner planets differ from outer planets in composition and structure?	Why is the study of planets important to understanding the universe and Earth's place in it?

MERCURY

History and Discovery

Mercury has been known since ancient times because it is one of the brightest objects in the night sky and can be seen without a telescope. The Babylonians, Greeks, and Romans all recorded observations of Mercury. Because it moves quickly across the sky, early astronomers once thought it was two different stars. Later, they realized it was a single planet orbiting very close to the Sun.

Composition & Physical Traits

Mercury is the smallest planet in the solar system, with a diameter of about 4,880 kilometers. It is a rocky, solid planet with a large metallic core. Mercury has no moons or rings. Mercury's weak gravity prevents it from holding a thick atmosphere, contributing to its heavily cratered surface from meteor collisions and extreme conditions. Daytime temperatures on Mercury can soar to 427°C (800°F), while nighttime temperatures plunge to around -180°C (-290°F).

Gravity

The surface gravity on Mercury is about 3.7 m/s^2, which is roughly 38% of Earth's gravity. This means you would weigh much less on Mercury than you do on Earth.

Location in the Solar System

Mercury is the closest planet to the Sun, orbiting at an average distance of about 58 million kilometers (36 million miles).

Viewing Mercury in the Sky

Mercury is difficult to see because it stays close to the Sun. It can be viewed shortly after sunset in the western sky or just before sunrise in the eastern sky. The best times to see Mercury are during its greatest elongations, when it appears farthest from the Sun as seen from Earth.

Name and Meaning

Mercury was named after the Roman messenger god, who was known for his speed. This name is fitting because Mercury orbits the Sun faster than any other planet, completing a full orbit in only 88 Earth days.

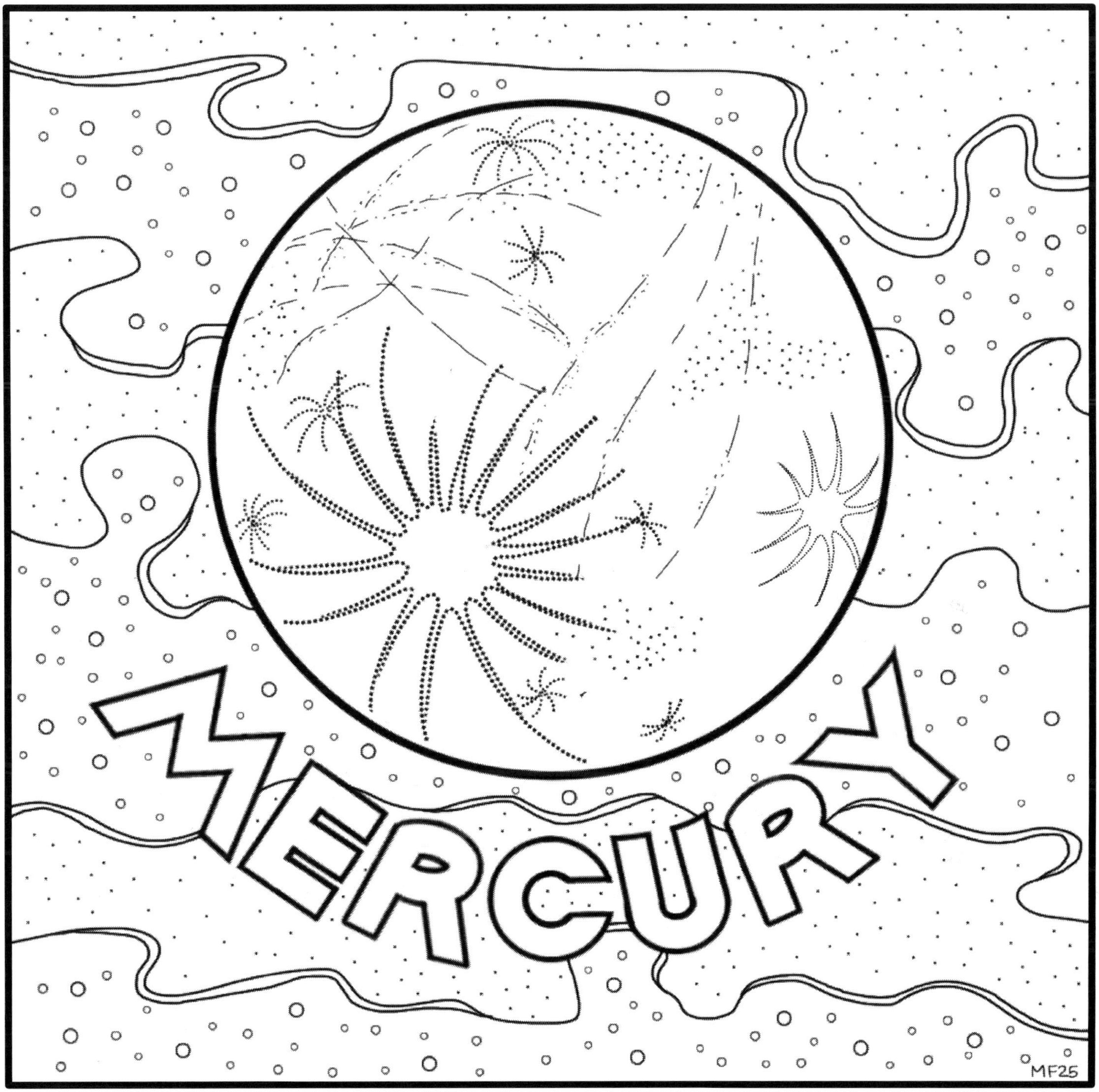

Name Origin:	Diameter:	State of Matter:
Interesting Facts:	Gravity:	Moons:
	Rings:	Km from sun:

VENUS

History and Discovery

Venus has been observed by humans since ancient times because it is one of the brightest objects in the sky, visible without a telescope. Ancient civilizations, including the Babylonians and Greeks, carefully tracked its movements. Early astronomers sometimes thought Venus was two different stars one appearing in the morning and one in the evening before realizing it was the same planet.

Composition & Physical Traits

Venus is similar in size to Earth, with a diameter of about 12,104 kilometers. Venus is a rocky planet with a dense atmosphere. Its surface is made of volcanic plains, mountains, and craters. The atmosphere is mostly carbon dioxide with thick clouds of sulfuric acid, which trap heat in a runaway greenhouse effect, making Venus the hottest planet in the solar system. Surface temperatures on Venus reach a blistering 460°C to 470°C (860°F to 878°F), which hot enough to melt lead. There is minimal variation between day and night temperature. Venus has no moons or rings.

Gravity

The surface gravity on Venus is about 8.87 m/s^2, which is roughly 90% of Earth's gravity. A person who weighs 100 pounds on Earth would weigh about 90 pounds on Venus.

Location in the Solar System

Venus is the second planet from the Sun, orbiting at an average distance of about 108 million kilometers (67 million miles).

Viewing Venus in the Sky

Venus is often called the "morning star" or the "evening star" because it is usually visible just before sunrise or just after sunset. It shines very brightly and is usually the third-brightest object in the sky after the Sun and the Moon.

Name and Meaning

Venus is named after the Roman goddess of love and beauty, the only planet named after a female deity. The name reflects the planet's striking brightness and beauty in the sky.

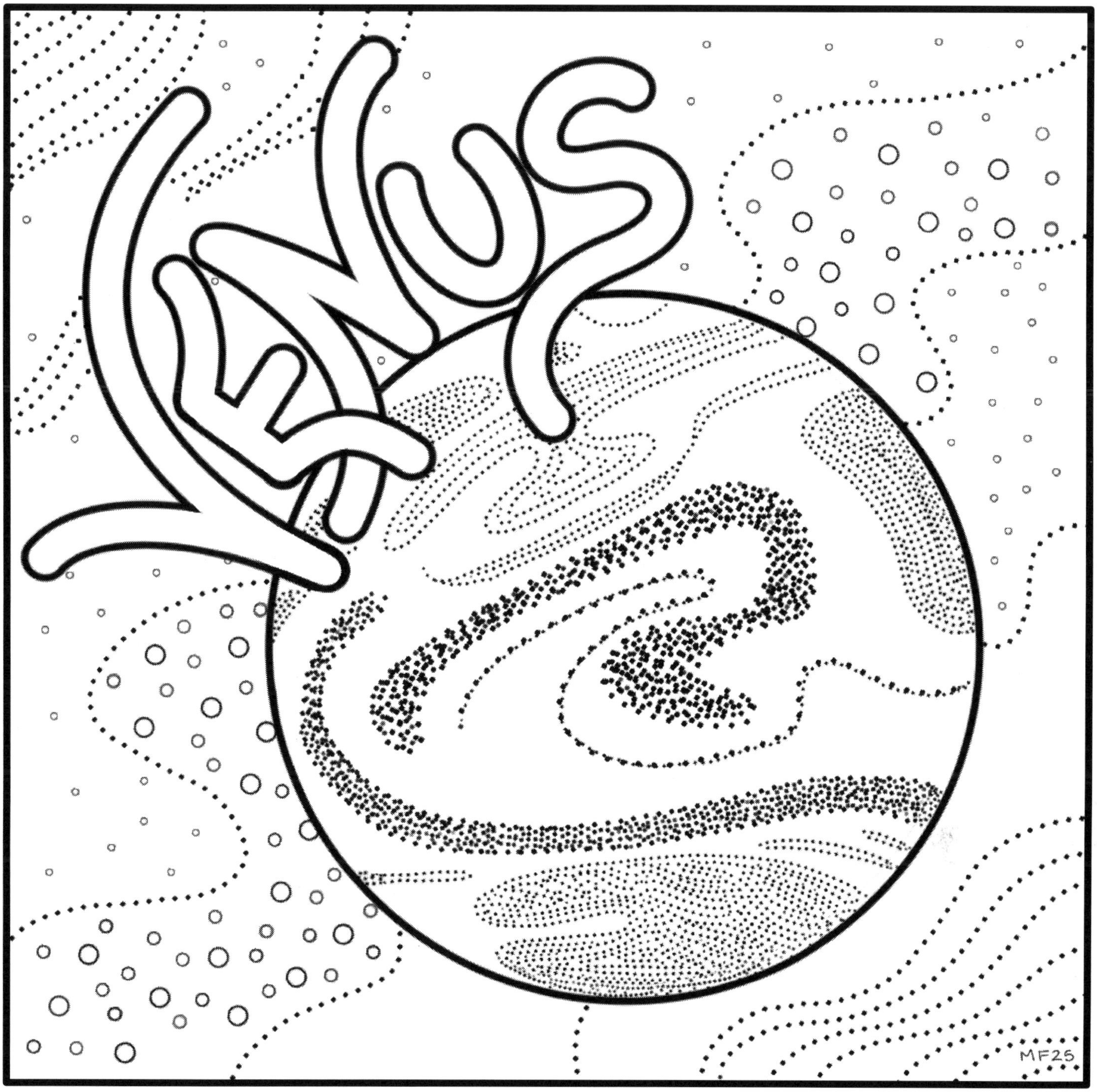

Name Origin:	Diameter:	State of Matter:
Interesting Facts:	Gravity:	Moons:
	Rings:	Km from sun:

EARTH

History and Discovery

Unlike other planets, Earth was not "discovered" because it is our home. Ancient cultures studied the skies and noticed that Earth was different from the "wandering stars" (planets). Over time, scientists such as Copernicus and Galileo helped show that Earth orbits the Sun, rather than being the center of the universe.

Composition & Physical Traits

Earth has a diameter of about 12,742 kilometers. Earth is a rocky planet made up of a solid inner core of iron and nickel, a liquid outer core, a thick mantle of hot rock, and a solid crust. About 71% of Earth's surface is covered by water, making it unique among the known planets. Earth is the only known planet where water can be found as solid, liquid, and gas. Earth also has a protective atmosphere composed mainly of nitrogen and oxygen, which supports life. Temperatures on Earth vary significantly, with officially recorded air temperatures ranging from as low as -89.2°C (-128.6°F) in Antarctica to as high as 56.7°C (134°F) in Death Valley, USA. Earth has one moon and no rings.

Gravity

Earth's surface gravity is about 9.8 m/s^2, which defines what we usually experience as "normal weight." This value is the standard used to compare gravity on other planets.

Location in the Solar System

Earth is the third planet from the Sun, orbiting at an average distance of about 150 million kilometers (93 million miles).

Viewing Earth in the Sky

From the surface, we cannot see Earth in the sky the way we see other planets, since we live on it. However, from space, Earth appears as a blue-and-white sphere due to its oceans and clouds.

Name and Meaning

The name "Earth" comes from Old English and Germanic words meaning ground or soil. Unlike other planets, Earth's name is not based on Roman or Greek mythology. Instead, it reflects its importance as the land beneath our feet and the planet that sustains life.

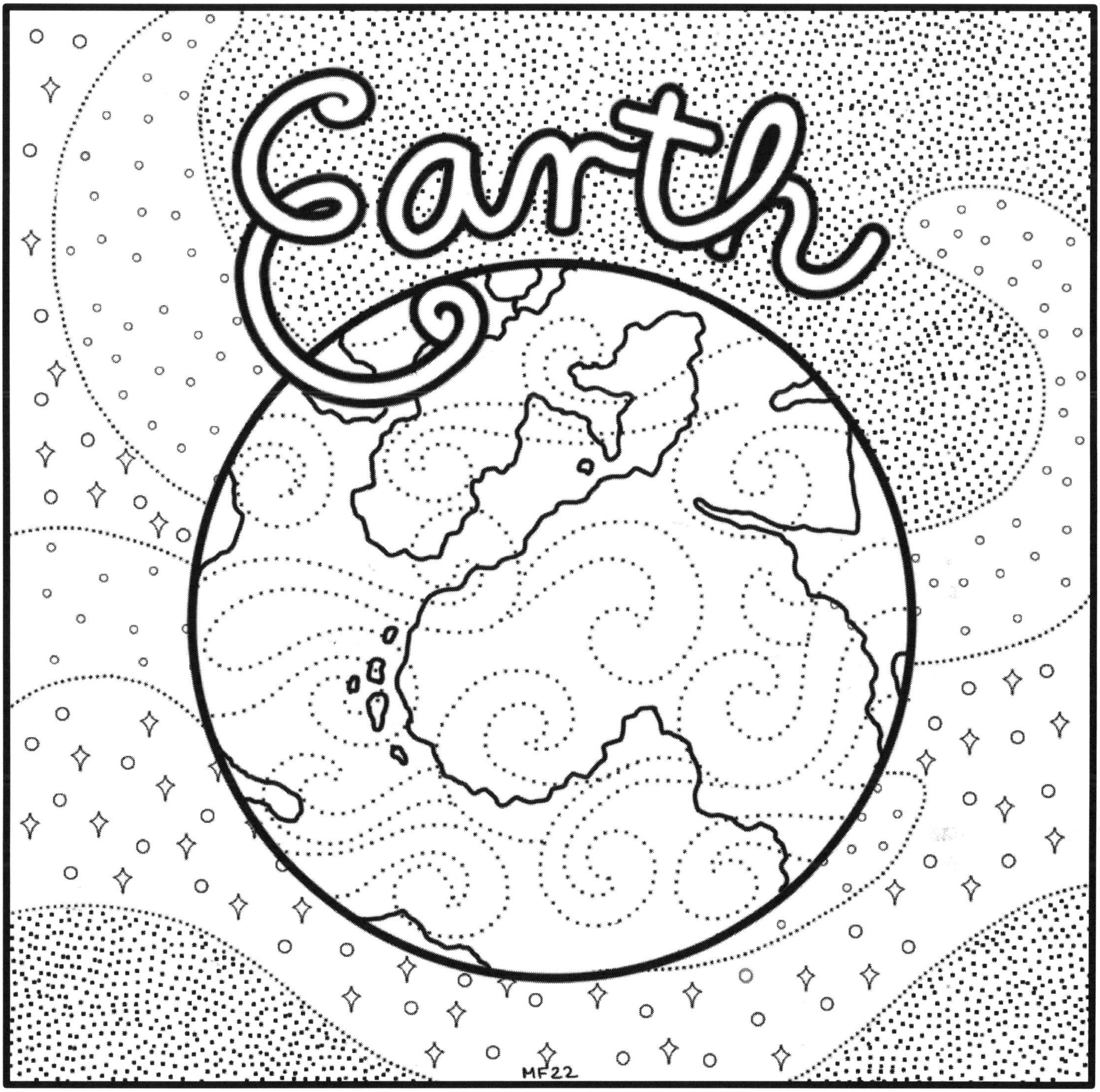

Name Origin:	Diameter:	State of Matter:

Interesting Facts:	Gravity:	Moons:
	Rings:	Km from sun:

MOON

The Moon is Earth's only natural satellite and has a diameter of about 3,474 kilometers, making it the fifth-largest moon in the solar system. It is a solid, rocky body with a surface made of rock, dust, and impact craters, and it has no rings. The Moon's surface gravity is approximately 1.62 m/s^2, which is about one-sixth of Earth's gravity, allowing objects to weigh much less than they do on Earth. On average, the Moon is about 384,400 kilometers from Earth and about 150 million kilometers from the Sun, since it travels with Earth as they orbit the Sun. The Moon takes about 27.3 days to complete one orbit around Earth and rotates at the same rate, which is why the same side of the Moon always faces Earth.

The Moon formed approximately 4.5 billion years ago, most likely from debris created when a Mars-sized object collided with early Earth. This material eventually came together under gravity to form the Moon. Humans have observed and studied the Moon since prehistoric times, using it to track time and seasons, but modern scientific understanding expanded dramatically during the Space Age, especially through robotic missions and the Apollo landings. The Moon plays a major role in Earth's ocean tides, as its gravitational pull causes water to rise and fall along coastlines. It also helps stabilize Earth's axial tilt, which contributes to relatively stable seasons over long periods of time. Historically, lunar phases have been used to guide agricultural planting cycles, calendars, and cultural traditions around the world.

The Moon's surface is made primarily of igneous rocks, including basalt, which formed from ancient lava flows, and anorthosite, a lighter-colored rock that makes up much of the Moon's highlands. The dark regions visible from Earth, called maria, are vast plains of hardened basalt created by volcanic activity billions of years ago. In contrast, the brighter highlands are older, heavily cratered regions made of anorthosite. The Moon's geology preserves a record of early solar system history because it lacks plate tectonics, erosion from water, and weathering processes that constantly reshape Earth's surface.

Unlike Earth, the Moon has no true atmosphere. Instead, it has an extremely thin layer of gases called an exosphere, made up of atoms such as helium, neon, and hydrogen. Because there is no thick atmosphere to trap heat or block radiation, temperatures on the Moon vary dramatically, ranging from extremely hot in sunlight to extremely cold in darkness. During the lunar day, temperatures on the Moon can soar to about 127°C (260°F). At night, temperatures can plunge to around –173°C (–280°F) since the Moon has no atmosphere to trap warmth, allowing heat to escape rapidly into space. The moon side facing Earth is called the near side, while the far side remains hidden from direct view due to the Moon's synchronized rotation, the moon is always facing Earth with the same face. The far side has fewer maria and a thicker crust, making it geologically different and one of the most mysterious regions of the Moon.

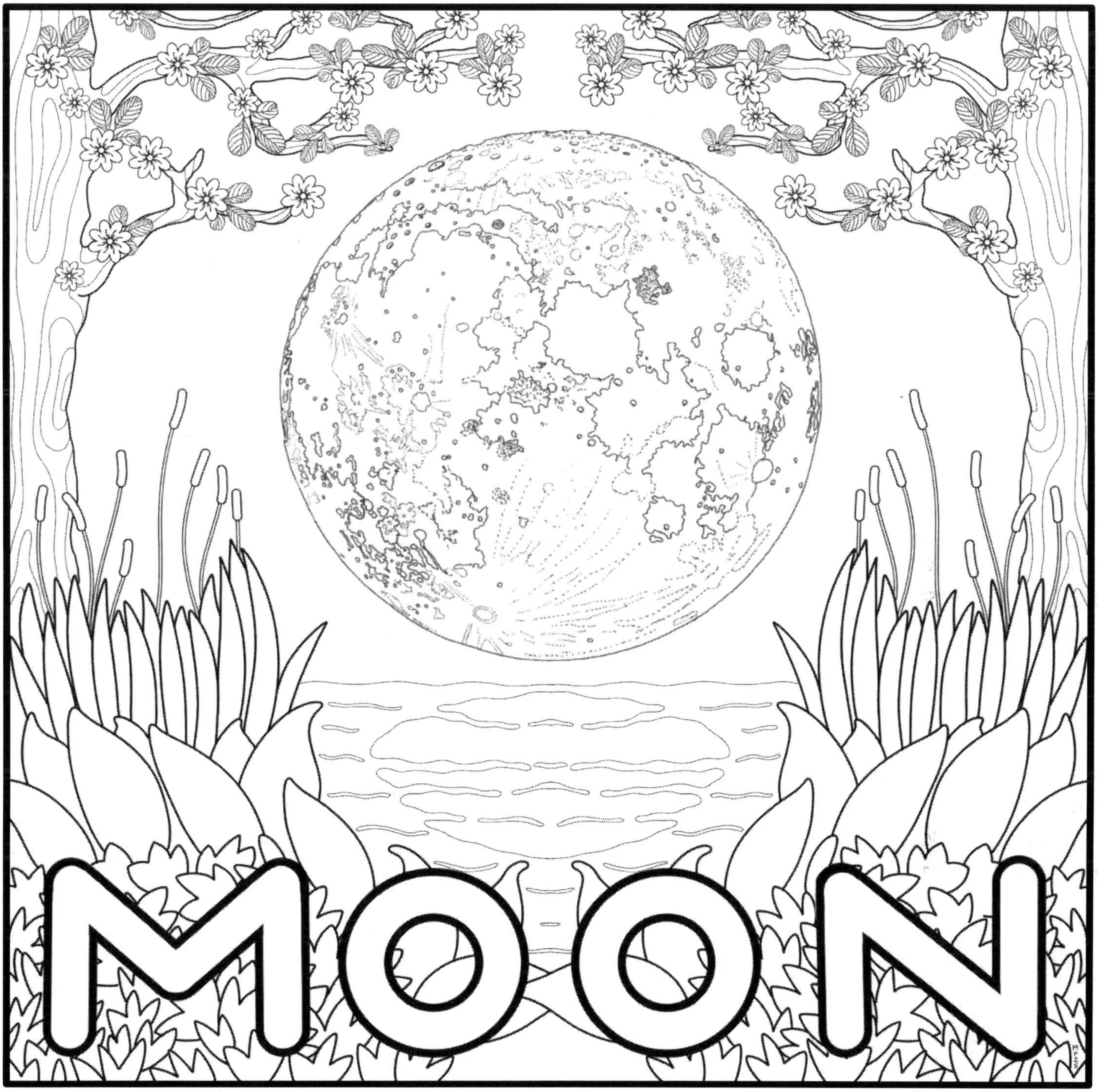

What are the Moon's basic physical characteristics, including its size, gravity, and orbit around Earth?	How did the Moon form, and what evidence supports this explanation?
How does the Moon influence Earth, both physically and historically?	What are the main differences between the Moon's maria and highlands.

LUNAR CYCLE

The Moon does not produce its own light; instead, it reflects sunlight. As the Moon orbits Earth, different portions of its sunlit half become visible from Earth. These changing views are known as the phases of the Moon. The complete cycle of lunar phases, called the lunar cycle, takes approximately 29.5 days and results from the relative positions of the Sun, Earth, and Moon.

The lunar cycle begins with the new moon, when the Moon is positioned between Earth and the Sun. During this phase, the side of the Moon facing Earth is not illuminated, making the Moon nearly invisible in the sky. As the Moon continues its orbit, a small portion of the sunlit side becomes visible, creating the waxing crescent phase. “Waxing” refers to the increasing amount of visible illumination. When half of the Moon’s disk is illuminated, the Moon reaches the first quarter phase, meaning one quarter of the lunar cycle has been completed.

Following the first quarter, the Moon enters the waxing gibbous phase, during which more than half of the Moon appears illuminated. The cycle then reaches the full moon, when Earth lies between the Sun and the Moon, allowing the entire sunlit side of the Moon to face Earth. After the full moon, the visible illuminated portion begins to decrease, a process known as waning. The Moon then passes through the waning gibbous phase and reaches the third (or last) quarter, where half of the Moon is again visible, but on the opposite side compared to the first quarter.

The final phase of the cycle is the waning crescent, when only a small portion of the Moon remains visible before it returns to the new moon phase and the cycle repeats. These phases occur in a predictable order and have been used throughout human history to track time, create calendars, and plan agricultural activities. The Moon’s phases also influence the timing and strength of ocean tides, with the strongest tides, known as spring tides, occurring during new and full moons when gravitational forces align.

A common misconception is that the Moon’s phases are caused by Earth’s shadow. In reality, Earth’s shadow only plays a role during lunar eclipses, which are rare events. The phases of the Moon are caused solely by changes in perspective as the Moon orbits Earth. Understanding lunar phases helps explain celestial motion, gravitational interactions, and the dynamic relationship between Earth, the Moon, and the Sun

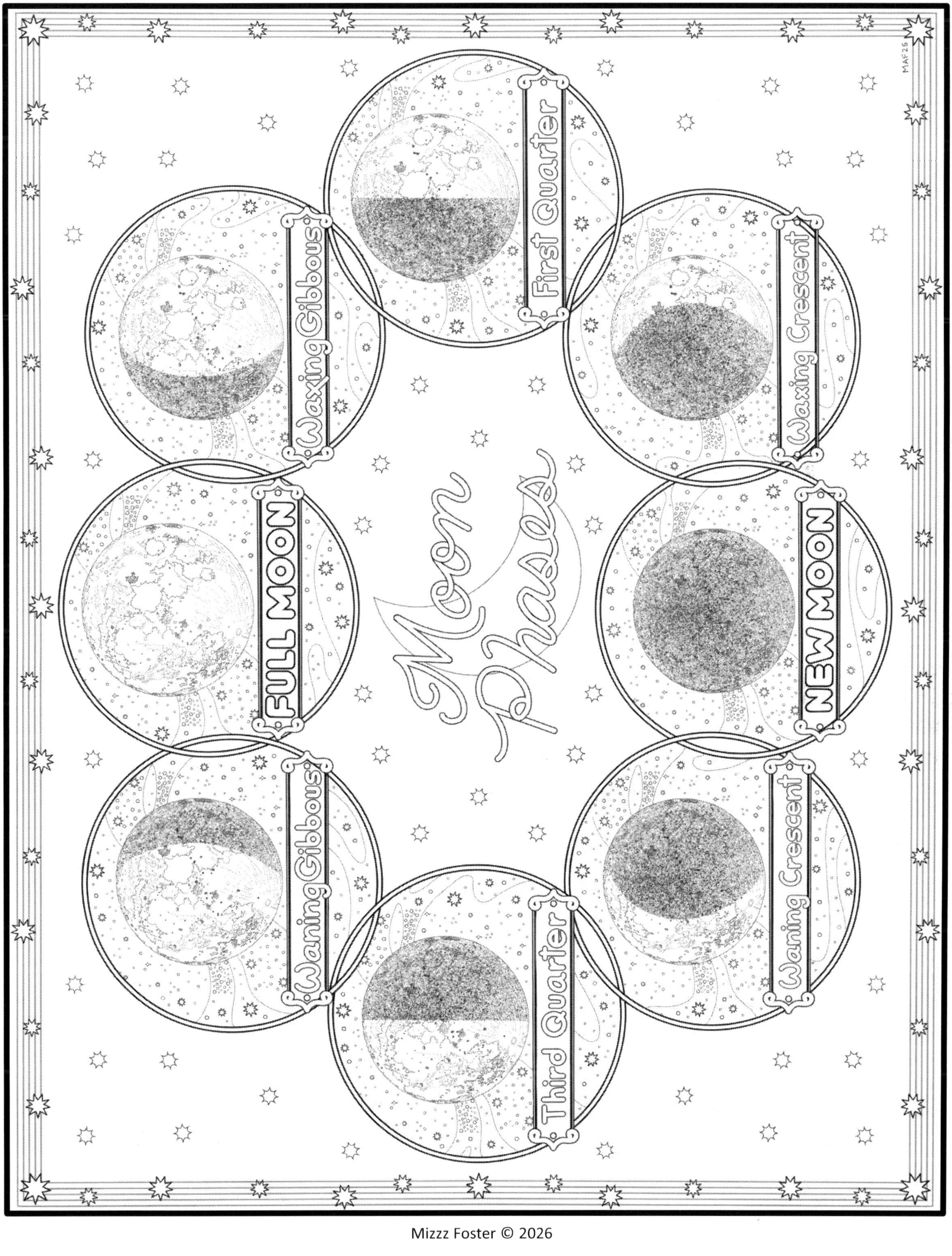
Moon Phases
First Quarter
Waxing Gibbous
Waxing Crescent
FULL MOON
NEW MOON
Waning Gibbous
Waning Crescent
Third Quarter
MAF25

ECLIPSE

An eclipse occurs when one celestial body moves into the shadow of another, temporarily blocking light from the Sun or altering how that light reaches Earth. In our solar system, eclipses involve the Sun, Earth, and the Moon. Eclipses are predictable astronomical events that provide important evidence about orbital motion, gravity, and the scale of the solar system.

There are two main types of eclipses observed from Earth: solar eclipses and lunar eclipses. A solar eclipse occurs when the Moon passes between the Sun and Earth, casting a shadow on Earth's surface. This can only happen during a new moon phase. Depending on the alignment and distances involved, a solar eclipse may be total, partial, or annular. In a total solar eclipse, the Moon completely blocks the Sun's disk, briefly turning day into darkness. In a partial eclipse, the Moon does not completely block the Sun's disk. In an annular eclipse, the Moon appears slightly smaller than the Sun, leaving a bright ring, or "annulus," of sunlight visible.

A lunar eclipse occurs when Earth moves between the Sun and the Moon, causing Earth's shadow to fall across the Moon. Lunar eclipses only occur during a full moon. Unlike solar eclipses, lunar eclipses are visible from anywhere on Earth where the Moon is above the horizon. During a total lunar eclipse, the Moon often appears reddish in color due to Earth's atmosphere scattering shorter wavelengths of light while allowing red wavelengths to reach the Moon, a phenomenon known as Rayleigh scattering.

Eclipse shadows are divided into regions called the umbra and penumbra. The umbra is the darkest part of the shadow, where the light source is completely blocked, while the penumbra is a lighter shadow where the light source is only partially blocked. The type and appearance of an eclipse depend on which part of the shadow reaches Earth or the Moon. These shadow regions help scientists understand the geometry and scale of eclipses.

Eclipses do not occur every month because the Moon's orbit is tilted about 5 degrees relative to Earth's orbital plane around the Sun. As a result, the Sun, Earth, and Moon usually do not align perfectly. Eclipses occur only during specific periods called eclipse seasons, which happen approximately twice a year when the Moon's orbit crosses Earth's orbital plane.

Historically, eclipses played a significant role in advancing scientific knowledge. Ancient civilizations recorded eclipses to refine calendars and predict celestial events. In modern science, eclipses have been used to test theories, such as confirming Einstein's theory of general relativity during a solar eclipse in 1919. Today, eclipses remain valuable opportunities for studying the Sun's atmosphere, the Moon's surface, and the dynamics of orbital motion.

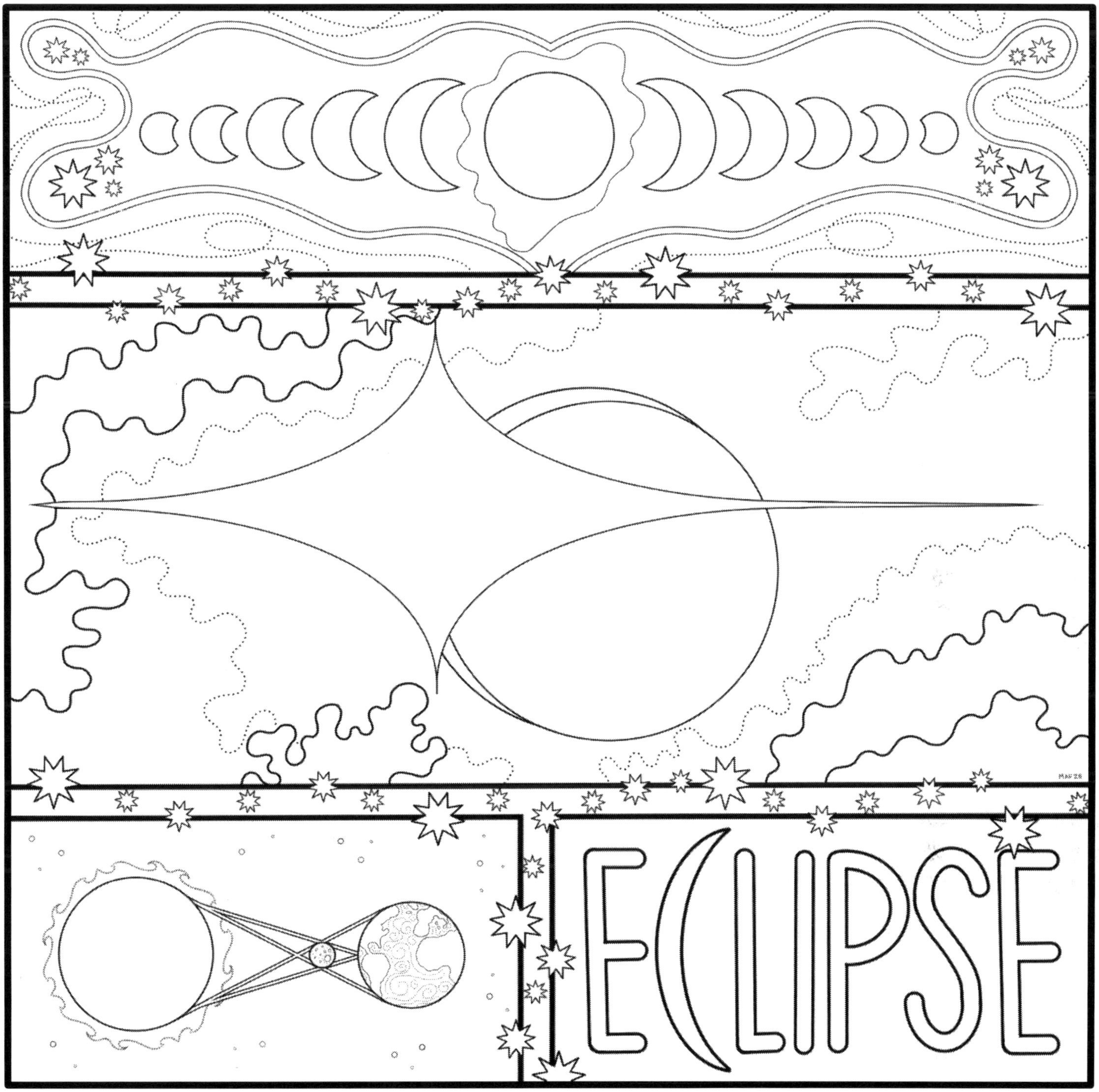

What is an eclipse, and which three celestial bodies are involved in eclipses observed from Earth?	How do solar eclipses and lunar eclipses differ in terms of alignment and moon phase?
What are the umbra and penumbra, and how do they affect the appearance of an eclipse?	Why do eclipses not occur every month?

MARS

History and Discovery

Mars has been observed in the night sky since ancient times because of its distinct red color. The Babylonians, Egyptians, and Greeks recorded its movements, and early astronomers noticed it wandered against the background of stars, classifying it as a planet. In the 1600s, Galileo was the first to view Mars through a telescope. Modern spacecraft such as the Viking landers and the Mars rovers have given us detailed information about the planet's surface and atmosphere.

Composition & Physical Traits

Mars has a diameter of approximately 6,779 kilometers. Mars is a rocky planet with a thin atmosphere made mostly of carbon dioxide. On Mars, daytime temperatures near the equator can rise to about 20°C (68°F) when the Sun is shining, but because of the very thin atmosphere, it cannot hold onto heat. At night, temperatures can drop dramatically to around –100°C (–148°F) or even colder. Its surface is covered with iron-rich dust and rocks, giving it the famous reddish appearance. Mars has the largest volcano in the solar system (Olympus Mons) and a massive canyon system (Valles Marineris). Scientists believe liquid water once flowed on Mars, and evidence suggests it may still exist as ice underground. Mars has two small moons, no rings.

Gravity

Mars has a surface gravity of about 3.7 m/s^2, which is only 38% of Earth's gravity. This means you would weigh much less on Mars than on Earth.

Location in the Solar System

Mars is the fourth planet from the Sun, orbiting at an average distance of about 228 million kilometers (142 million miles).

Viewing Mars in the Sky

Mars is visible to the naked eye and appears as a reddish "star" in the night sky. The best time to view Mars is during opposition, which occurs about every two years when Earth is directly between Mars and the Sun, making Mars appear brighter and closer.

Name and Meaning

Mars is named after the Roman god of war, because of its red color, which reminded ancient people of blood and battle. Its Greek counterpart was Ares. The name reflects its striking and fiery appearance in the night sky.

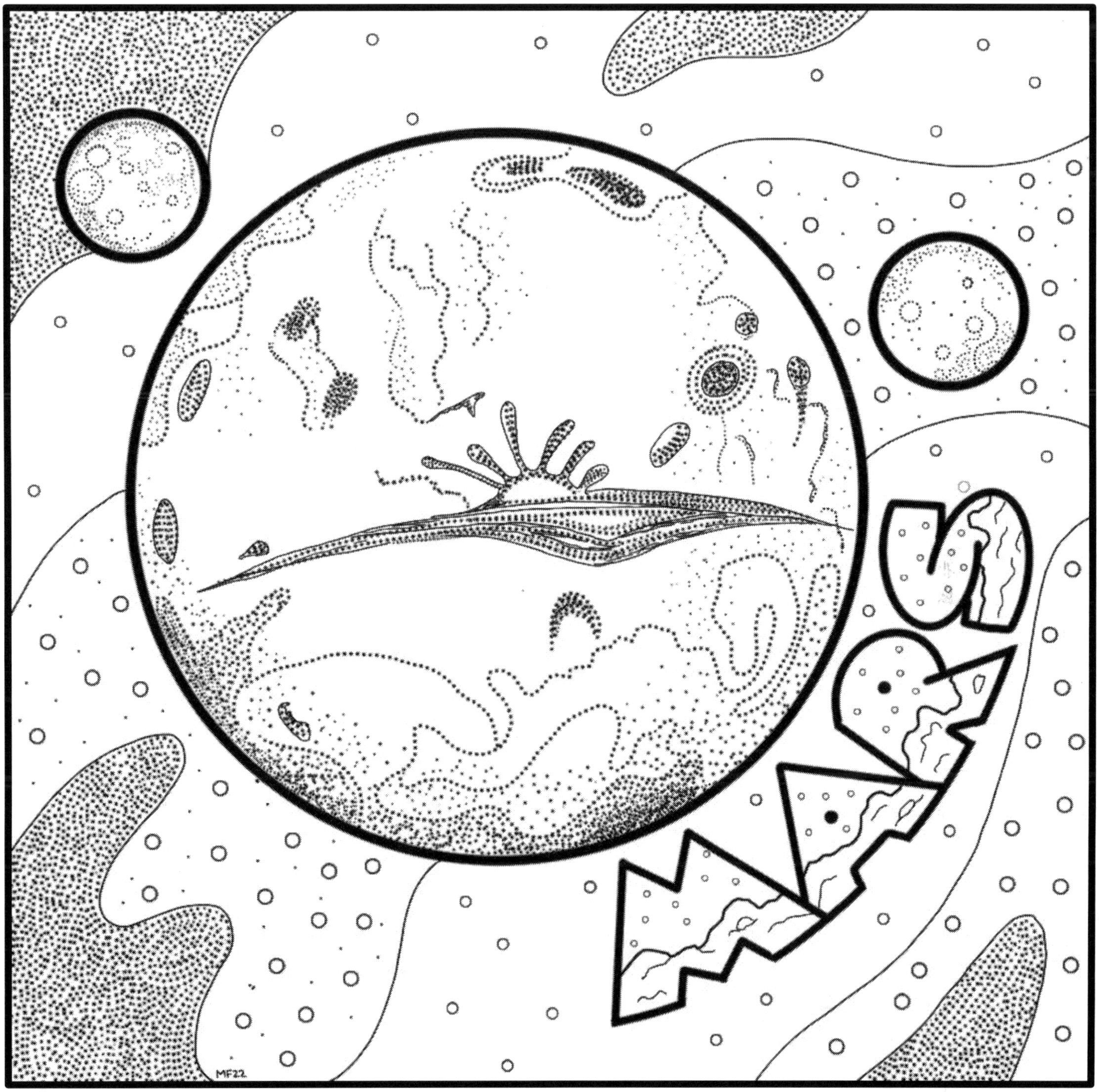

Name Origin:	
Interesting Facts:	

Diameter:	State of Matter:
Gravity:	Moons:
Rings:	Km from sun:

ASTEROID BELT

The asteroid belt is a large region of the solar system located between the orbits of Mars and Jupiter, extending from approximately 329 million to 478 million kilometers from the Sun. Although it is often shown as a dense band in illustrations, the asteroid belt is very spread out, with most asteroids separated by vast distances. The belt is relatively thin vertically, with asteroid orbits lying close to the same flat plane as the planets, known as the ecliptic plane.

The asteroid belt is composed of millions of rocky and metallic objects, ranging in size from microscopic dust to large bodies hundreds of kilometers across. The largest object in the belt is Ceres, which is classified as a dwarf planet. Most asteroids are made of rock, metal, or a mixture of rock and ice, and many are considered leftover building material from the early solar system.

Scientists believe the asteroid belt formed about 4.6 billion years ago during the formation of the planets. Early in the solar system's history, material in this region could not combine into a planet because of the strong gravitational influence of Jupiter, which disrupted accretion by increasing collision speeds and preventing objects from sticking together.

The asteroid belt was first recognized in the early 19th century, beginning with the discovery of Ceres in 1801. As additional objects were discovered in similar orbits, astronomers realized they were part of a continuous population rather than individual planets. Today, spacecraft missions and telescopic observations continue to expand scientific understanding of this region.

The asteroid belt is important because it preserves primitive material from the early solar system. Studying asteroids provides insight into planet formation, solar system evolution, and the origins of meteorites that sometimes reach Earth. Despite its dramatic portrayal in science fiction, the asteroid belt is not dangerous to spacecraft and is a valuable region for scientific exploration.

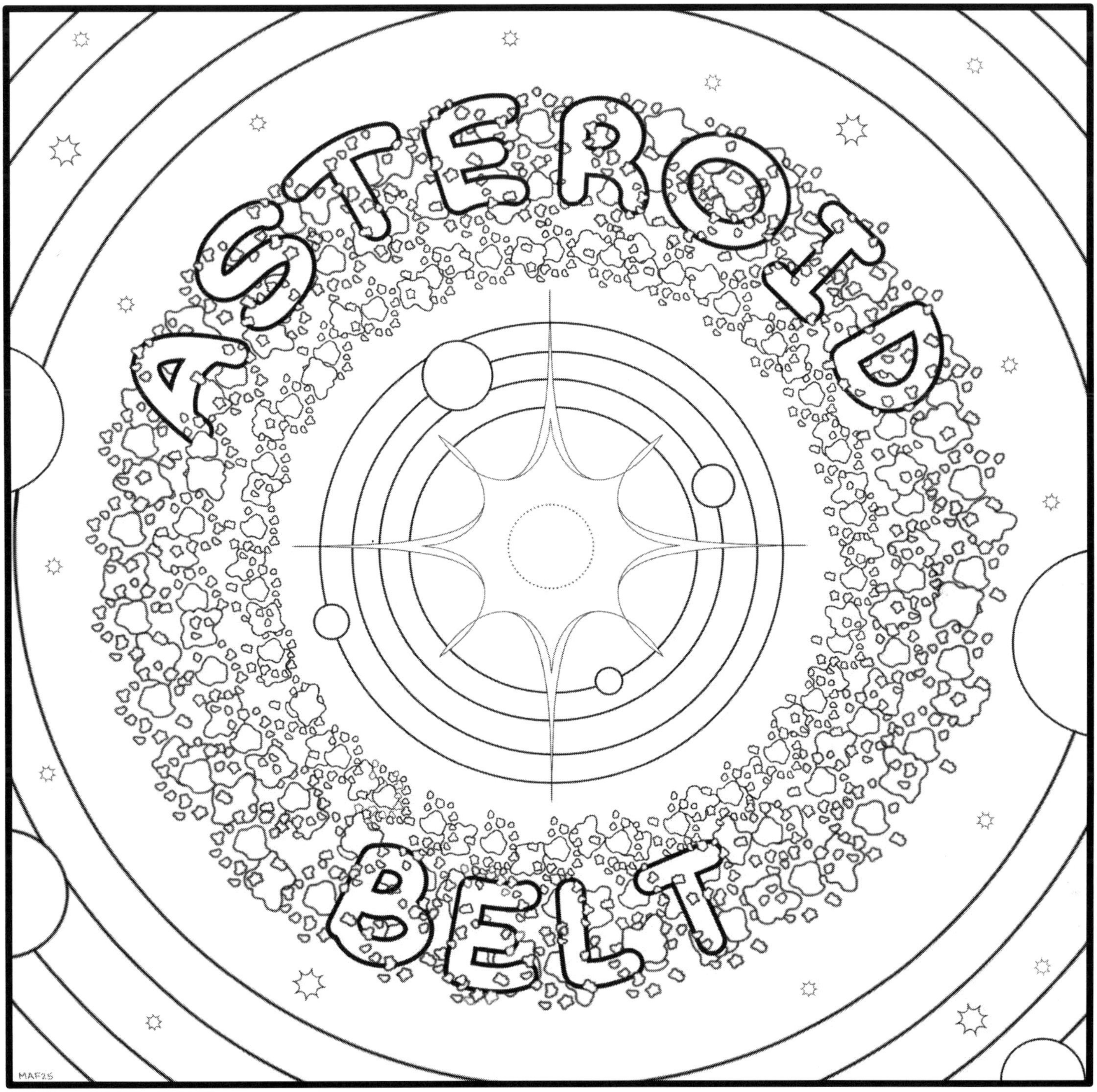

Where is the asteroid belt located in the solar system, and how does its actual structure differ from how it is often shown in illustrations?	What types of materials make up most asteroids, and what is the significance of Ceres within the asteroid belt?
Why was a planet unable to form in the region of the asteroid belt, and what role did Jupiter play in this process?	Why is the asteroid belt important to scientists studying the early solar system and meteorites?

JUPITER

History and Discovery

Jupiter has been visible in the night sky since ancient times and was observed by civilizations such as the Babylonians, Egyptians, and Greeks. In 1610, Galileo Galilei used a telescope to discover Jupiter's four largest moons—Io, Europa, Ganymede, and Callisto—now called the Galilean moons. These discoveries provided strong evidence that not everything revolved around Earth.

Composition & Physical Traits

Jupiter is a gas giant, composed mostly of hydrogen and helium, similar to the Sun. Jupiter is the largest planet in the solar system, with a diameter of about 139,820 kilometers. It has swirling clouds of ammonia and water in its atmosphere, along with the famous Great Red Spot, a giant storm that has lasted for centuries. Jupiter may have a small rocky core deep inside, surrounded by layers of liquid metallic hydrogen. Jupiter has ninety-five moons, a faint ring system. Jupiter does not have a solid surface like Earth, and temperatures are measured at the tops of its clouds. At cloud level, temperatures are extremely cold, averaging about –145°C (–234°F), and there is little difference between day and night because Jupiter's thick atmosphere and internal heat help distribute energy around the planet.

Gravity

Jupiter has the strongest gravity of any planet in the solar system, about 24.8 m/s^2, which is 2.5 times stronger than Earth's gravity. Although Jupiter is massive, you couldn't stand on it because it has no solid surface.

Location in the Solar System

Jupiter is the fifth planet from the Sun, orbiting at an average distance of about 778 million kilometers (484 million miles).

Viewing Jupiter in the Sky

Jupiter is often one of the brightest objects in the night sky and can usually be seen with the naked eye. With binoculars or a small telescope, observers can even spot its four largest moons; Io, Europa, Ganymede, and Callisto. They are called the Galilean moons because they were discovered in 1610 by Galileo Galilei. The best times to view Jupiter are during opposition, which happens about every 13 months, when Earth passes directly between Jupiter and the Sun.

Name and Meaning

Jupiter is named after the king of the Roman gods, the counterpart of the Greek god Zeus. The name reflects its enormous size and importance, as it is the largest planet in the solar system.

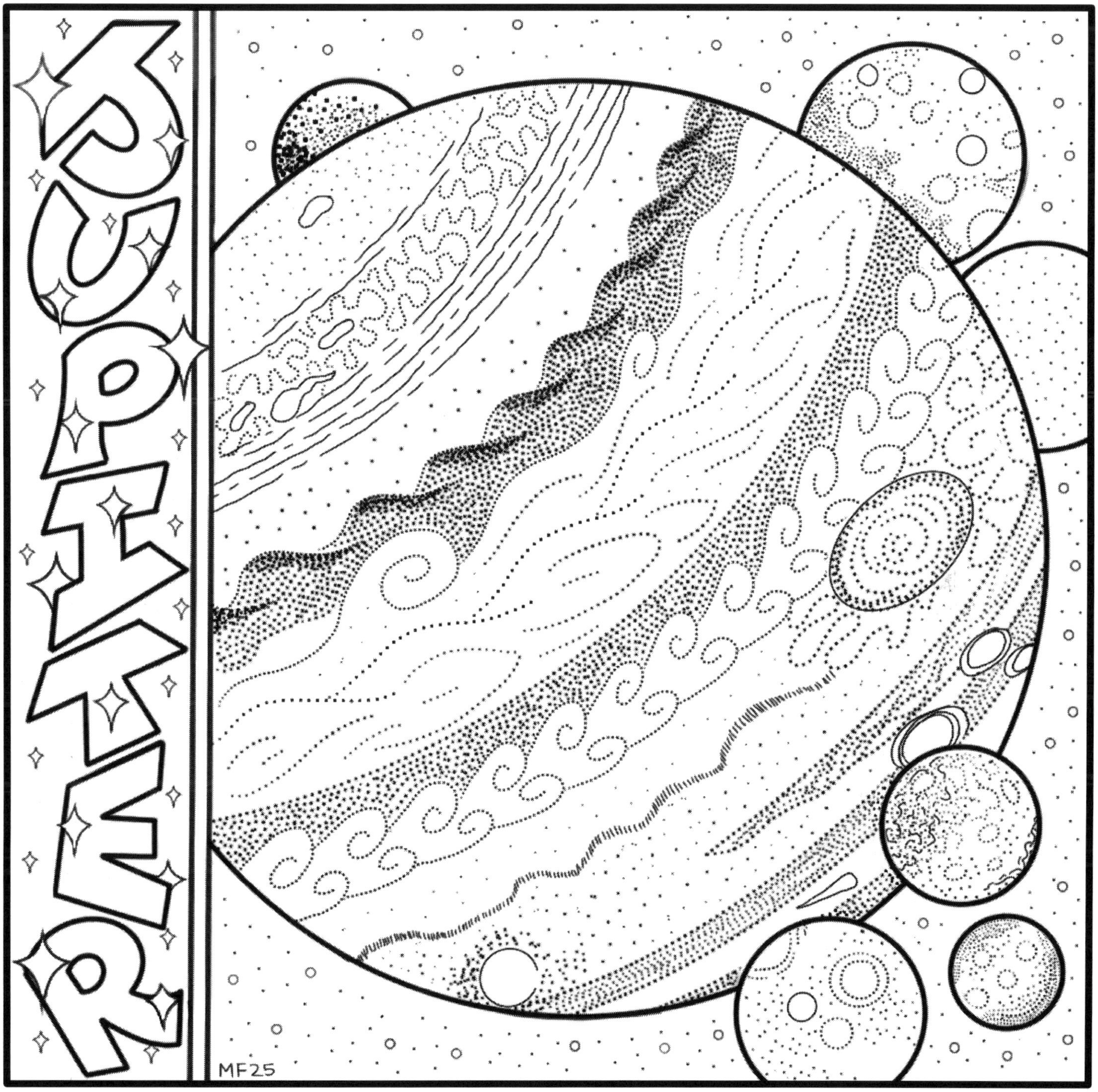

Name Origin:

Diameter:

State of Matter:

Interesting Facts:

Gravity:

Moons:

Rings:

Km from sun:

GALILEAN MOONS

The Galilean moons are the four largest moons of Jupiter: Io, Europa, Ganymede, and Callisto. They are called the Galilean moons because they were discovered in 1610 by Galileo Galilei using one of the first astronomical telescopes. This discovery was historically important because it provided strong evidence that not all objects orbit Earth, supporting the heliocentric model of the solar system.

Together, the Galilean moons form a miniature planetary system. They orbit Jupiter in orderly paths and are locked in a gravitational relationship called orbital resonance, which strongly affects their geology and internal structure. These moons are among the most scientifically important objects in the solar system due to their size, diversity, and potential to host conditions suitable for life.

Io

Io is the innermost Galilean moon and the most volcanically active body in the solar system. It has a diameter of approximately 3,643 kilometers, making it slightly larger than Earth's Moon. Io's intense volcanism is caused by tidal heating, which occurs as Jupiter's powerful gravity stretches and squeezes the moon. Its surface is covered with sulfur and lava flows, giving it bright yellow, orange, and red colors. Io has hundreds of active volcanoes, some of which erupt lava fountains hundreds of kilometers high.

Europa

Europa has a diameter of about 3,122 kilometers and is slightly smaller than Earth's Moon. It is covered by a thick shell of ice, beneath which scientists believe lies a global subsurface ocean of liquid water. This ocean is kept warm by tidal heating from Jupiter's gravity. Europa's surface is smooth and crisscrossed with cracks and ridges, suggesting active geological processes. Because liquid water is essential for life, Europa is considered one of the best places to search for life beyond Earth.

Ganymede

Ganymede is the largest moon in the solar system, with a diameter of approximately 5,268 kilometers, making it larger than the planet Mercury. It is unique because it is the only moon known to have its own magnetic field. Ganymede has a layered interior, with a metallic core, rocky mantle, and icy outer shell. Evidence suggests it may also contain a subsurface ocean, making it another important target for future exploration.

Callisto

Callisto has a diameter of about 4,821 kilometers and is the outermost Galilean moon. It has one of the oldest and most heavily cratered surfaces in the solar system, showing little geological activity. Unlike Io and Europa, Callisto experiences minimal tidal heating. Scientists believe Callisto may also harbor a subsurface ocean beneath its icy crust. Its stable surface and lower radiation levels make it a potential candidate for future space missions or human exploration.

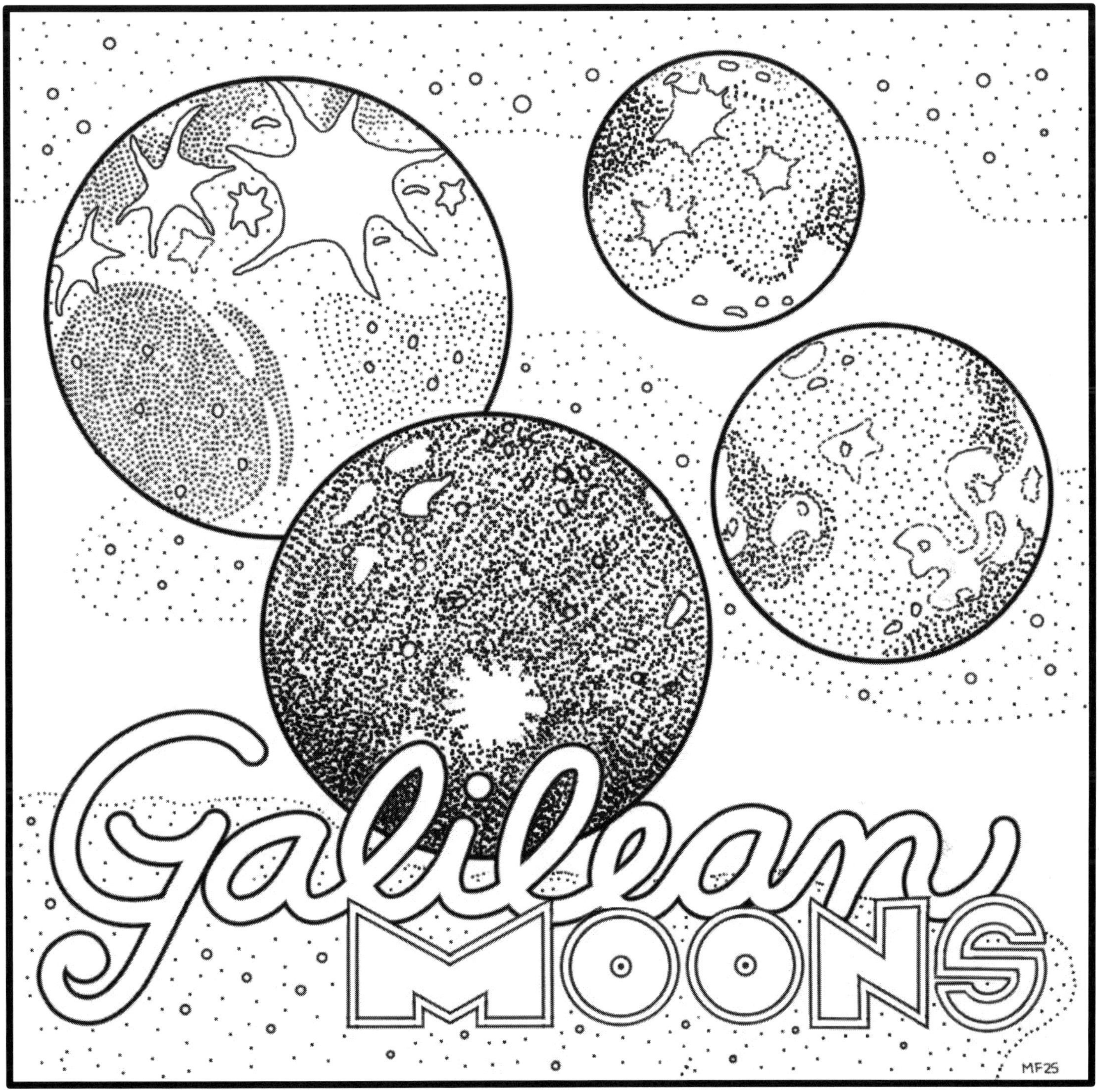

What are the names of the four Galilean moons and which planet do they orbit?

Who discovered the Galilean moons, and why was the discovery important?

What makes Ganymede unique?

Why do scientists think the Galilean moons might support life?

SATURN

History and Discovery

Saturn has been known since ancient times because it can be seen with the naked eye. Ancient cultures including the Babylonians and Romans recorded its movements across the sky. In 1610, Galileo Galilei was the first to observe Saturn through a telescope. He saw the planet's unusual "ears," which we now know were its rings. Later, astronomer Christiaan Huygens correctly identified the rings and discovered Saturn's largest moon, Titan.

Composition & Physical Traits

Saturn is a gas giant made mostly of hydrogen and helium. Saturn is a gas giant, so temperatures are measured at the tops of its clouds rather than on a solid surface. At cloud level, temperatures average about –178°C (–288°F), and there is very little difference between day and night because Saturn's thick atmosphere and internal heat help distribute energy evenly around the planet. Saturn has a diameter of about 116,460 kilometers. Beneath Saturn's thick atmosphere, there may be a small rocky and icy core surrounded by liquid metallic hydrogen. Saturn's most famous feature is its stunning system of rings, made of billions of ice and rock particles ranging from tiny grains to house-sized chunks. Saturn has 274 moons, more than any other planet in the solar system.

Gravity

The surface gravity on Saturn is about 10.4 m/s^2, only slightly stronger than Earth's gravity, even though Saturn is much larger. This is because Saturn has a low density, it is the least dense planet in the solar system and would float in water if there were an ocean big enough!

Location in the Solar System

Saturn is the sixth planet from the Sun, orbiting at an average distance of about 1.4 billion kilometers (886 million miles).

Viewing Saturn in the Sky

Saturn is visible to the naked eye as a bright, steady point of light. With a telescope, observers can see its rings and some of its moons. The best times to view Saturn are during opposition, which occurs about every 12 months, when Saturn is closest to Earth and fully illuminated by the Sun.

Name and Meaning

Saturn is named after the Roman god of agriculture and time, who was also the father of Jupiter (Zeus in Greek mythology). The name reflects the planet's importance and its slow movement across the sky, as Saturn takes nearly 30 Earth years to complete one orbit around the Sun.

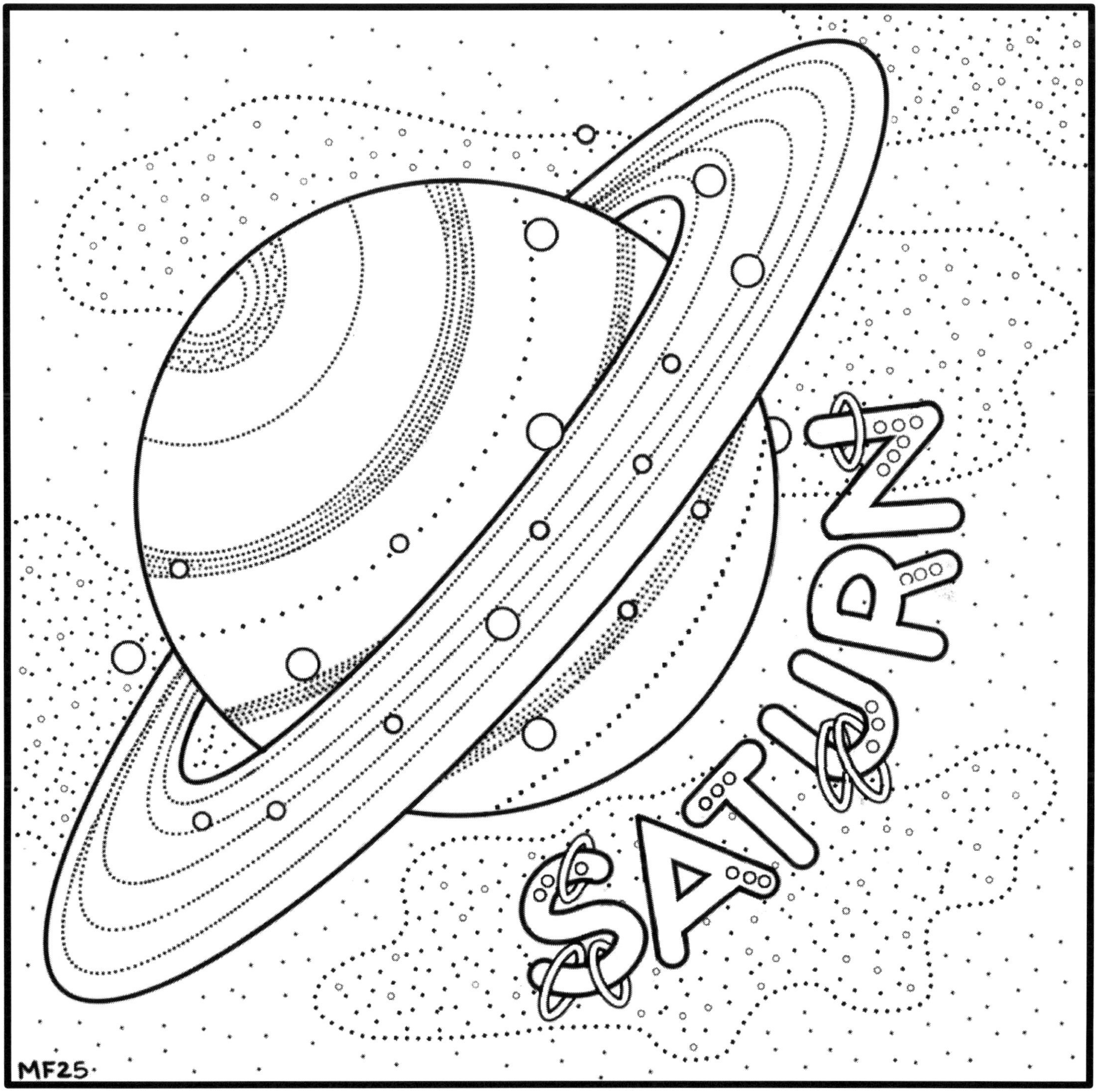

Name Origin:	Diameter:	State of Matter:
Interesting Facts:	Gravity:	Moons:
	Rings:	Km from sun:

URANUS

History and Discovery

Unlike the planets visible to the naked eye, Uranus was not officially recognized until the invention of the telescope. In 1781, astronomer William Herschel discovered Uranus while scanning the night sky. At first, he thought it was a star, but after careful observations, he realized it was a new planet, the first one discovered in modern history.

Composition & Physical Traits

Uranus has a diameter of approximately 50,724 kilometers. Uranus is an ice giant, made mostly of hydrogen and helium, with large amounts of water, ammonia, and methane. The methane gas in its atmosphere absorbs red light and reflects blue-green light, giving Uranus its distinctive color. Beneath the atmosphere lies a mantle of icy materials and possibly a rocky core. Its extreme tilt causes unusual seasons. Uranus is an ice giant, and temperatures are measured at the tops of its clouds rather than on a solid surface. At cloud level, temperatures average about –224°C (–371°F), making it one of the coldest planets in the solar system. Because Uranus has a thick atmosphere and receives very little sunlight, there is only a small difference between daytime and nighttime temperatures. Uranus has twenty-eight moons, and a faint ring system.

Gravity

The surface gravity on Uranus is about 8.7 m/s^2, which is very similar to that of Earth and Venus. Even though Uranus is much larger than Earth, Uranus's low density makes its gravity slightly weaker.

Location in the Solar System

Uranus is the seventh planet from the Sun, orbiting at an average distance of about 2.9 billion kilometers (1.8 billion miles). It is unique because it rotates on its side, tilted by about 98 degrees, likely due to a massive collision in its past.

Viewing Uranus in the Sky

Uranus is sometimes just visible to the naked eye under very dark skies, but it usually requires binoculars or a telescope to see clearly. It appears as a small, bluish-green disk. The best time to view Uranus is during opposition, which occurs once a year when Earth passes directly between Uranus and the Sun.

Name and Meaning

Uranus was named after the ancient Greek god of the sky, who was the father of Cronus (Saturn) and grandfather of Zeus (Jupiter). It is the only planet named after a Greek god rather than a Roman one, reflecting its connection to the heavens.

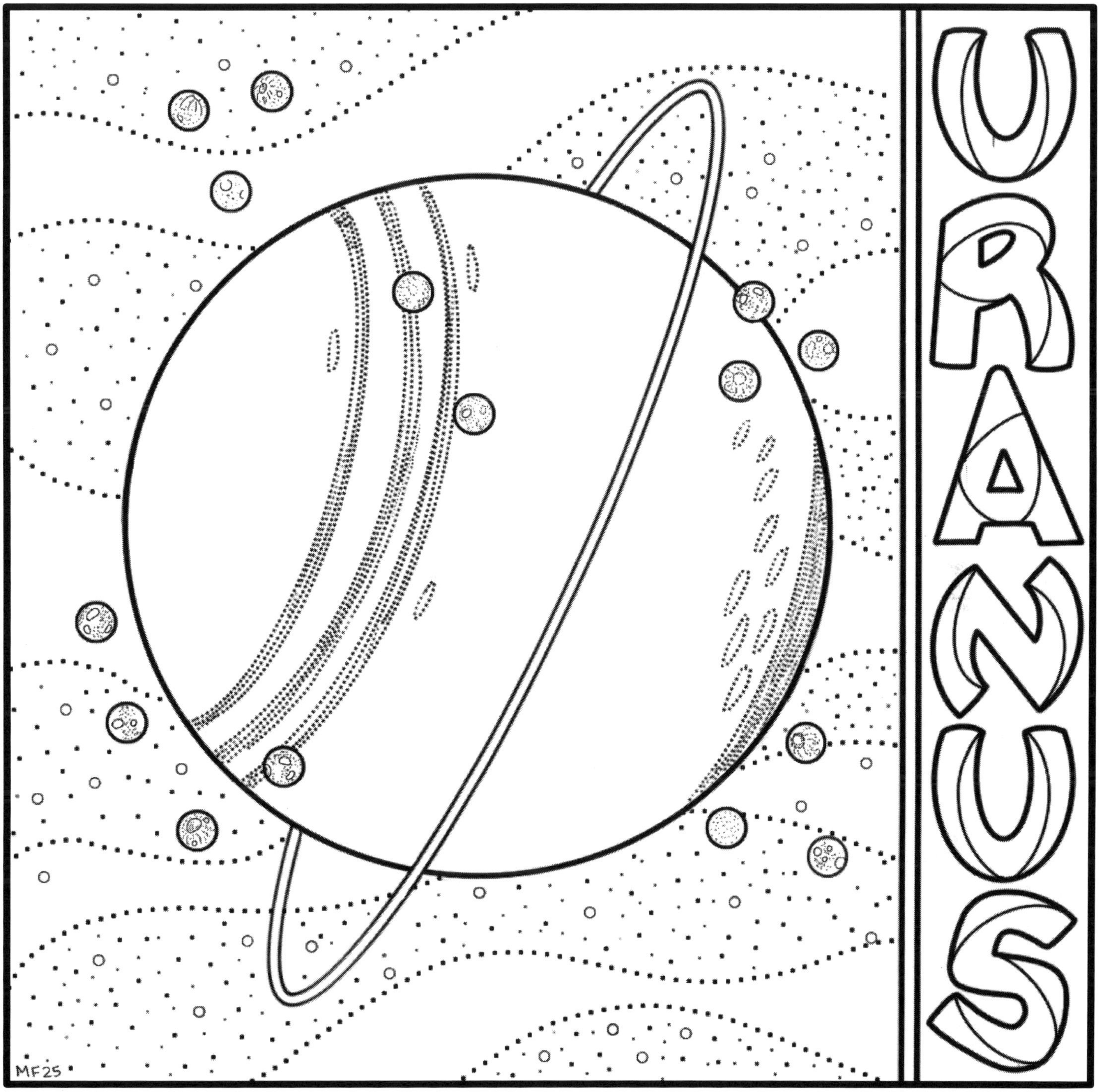

Name Origin:	Diameter:	State of Matter:
Interesting Facts:	Gravity:	Moons:
	Rings:	Km from sun:

NEPTUNE

History and Discovery

Neptune was the first planet discovered through mathematical prediction rather than direct observation. In the mid-1800s, astronomers noticed that Uranus's orbit was slightly irregular. French mathematician Urbain Le Verrier and English mathematician John Couch Adams independently calculated where an unknown planet should be. In 1846, German astronomer Johann Galle confirmed their predictions and officially discovered Neptune with a telescope.

Composition & Physical Traits

Neptune is an ice giant, made mostly of hydrogen, helium, water, ammonia, and methane. Neptune has a diameter of about 49,244 kilometers. Its methane-rich atmosphere gives the planet a deep blue color. Neptune also has powerful winds—the fastest in the solar system—reaching speeds over 2,000 km/h, along with large storms like the Great Dark Spot. Neptune has 16 known moons and faint rings. Neptune is an ice giant, and temperatures are measured at the tops of its clouds rather than on a solid surface. At cloud level, temperatures average about –214°C (–353°F), and there is little difference between day and night because Neptune's thick atmosphere and internal heat help circulate energy throughout the planet.

Gravity

Neptune's surface gravity is about 11.0 m/s^2, a little stronger than Earth's. Although Neptune is smaller than Jupiter and Saturn, its higher density makes its gravity fairly strong.

Location in the Solar System

Neptune is the eighth and farthest planet from the Sun, orbiting at an average distance of about 4.5 billion kilometers (2.8 billion miles). It takes about 165 Earth years for Neptune to complete one orbit.

Viewing Neptune in the Sky

Neptune is too faint to be seen with the naked eye. It requires binoculars or a telescope to observe and appears as a small, bluish dot in the sky. The best time to view Neptune is during opposition, which happens once a year when Earth is directly between Neptune and the Sun.

Name and Meaning

Neptune was named after the Roman god of the sea, known as Poseidon in Greek mythology. The name reflects the planet's striking blue color, reminiscent of Earth's oceans.

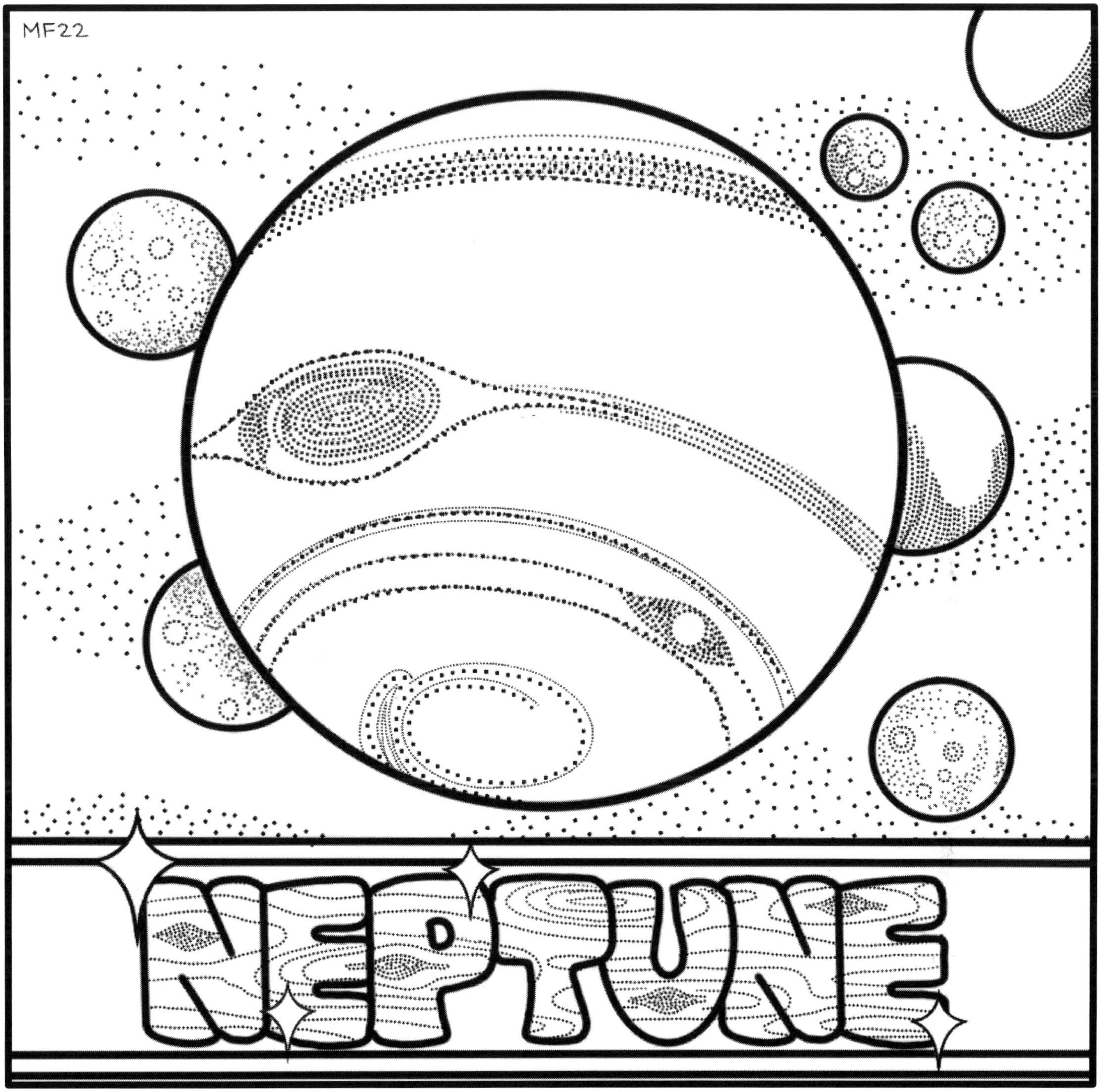

Name Origin:	Diameter:
	State of Matter:
Interesting Facts:	Gravity:
	Moons:
	Rings:
	Km from sun:

PLUTO

History and Discovery

Pluto was discovered in 1930 by American astronomer Clyde Tombaugh at the Lowell Observatory in Arizona. Its existence was predicted because astronomers noticed unusual patterns in the orbits of Neptune and Uranus. For many years, Pluto was considered the ninth planet in the solar system. In 2006, however, the International Astronomical Union reclassified Pluto as a dwarf planet because it does not clear other objects from its orbital path.

Composition & Physical Traits

Pluto is made of a mixture of rock and ice and has a diameter of about 2,377 km. Its surface is covered with frozen nitrogen, methane, and carbon monoxide, with mountains likely made of water ice. Pluto also has a thin atmosphere that expands when the planet is closer to the Sun and freezes when it moves farther away. Pluto is extremely cold because it is so far from the Sun. During the day, temperatures can reach about –218°C (–360°F), while at night they can drop to around –240°C (–400°F). Pluto's thin, temporary atmosphere freezes and falls to the surface at night, contributing to its dramatic and frigid temperature changes. Pluto has five moons and no known rings.

Gravity

Pluto's surface gravity is about 0.62 m/s^2, which is only about 6% of Earth's gravity. This means a person weighing 100 pounds on Earth would weigh just 6 pounds on Pluto.

Location in the Solar System

Pluto is located in the Kuiper Belt, a region beyond Neptune filled with icy bodies and dwarf planets. It orbits the Sun at an average distance of about 5.9 billion kilometers (3.7 billion miles). One Pluto year equals about 248 Earth years.

Viewing Pluto in the Sky

Pluto is very faint and cannot be seen with the naked eye. It requires a powerful telescope to observe, appearing only as a tiny point of light.

Name and Meaning

Pluto was named after the Roman god of the underworld, chosen in part because of its distant, dark, and cold environment. The name was suggested by an 11-year-old English schoolgirl, Venetia Burney. The planet's symbol, ♇, combines the letters P and L, which also honor Percival Lowell, the astronomer whose work led to Pluto's discovery.

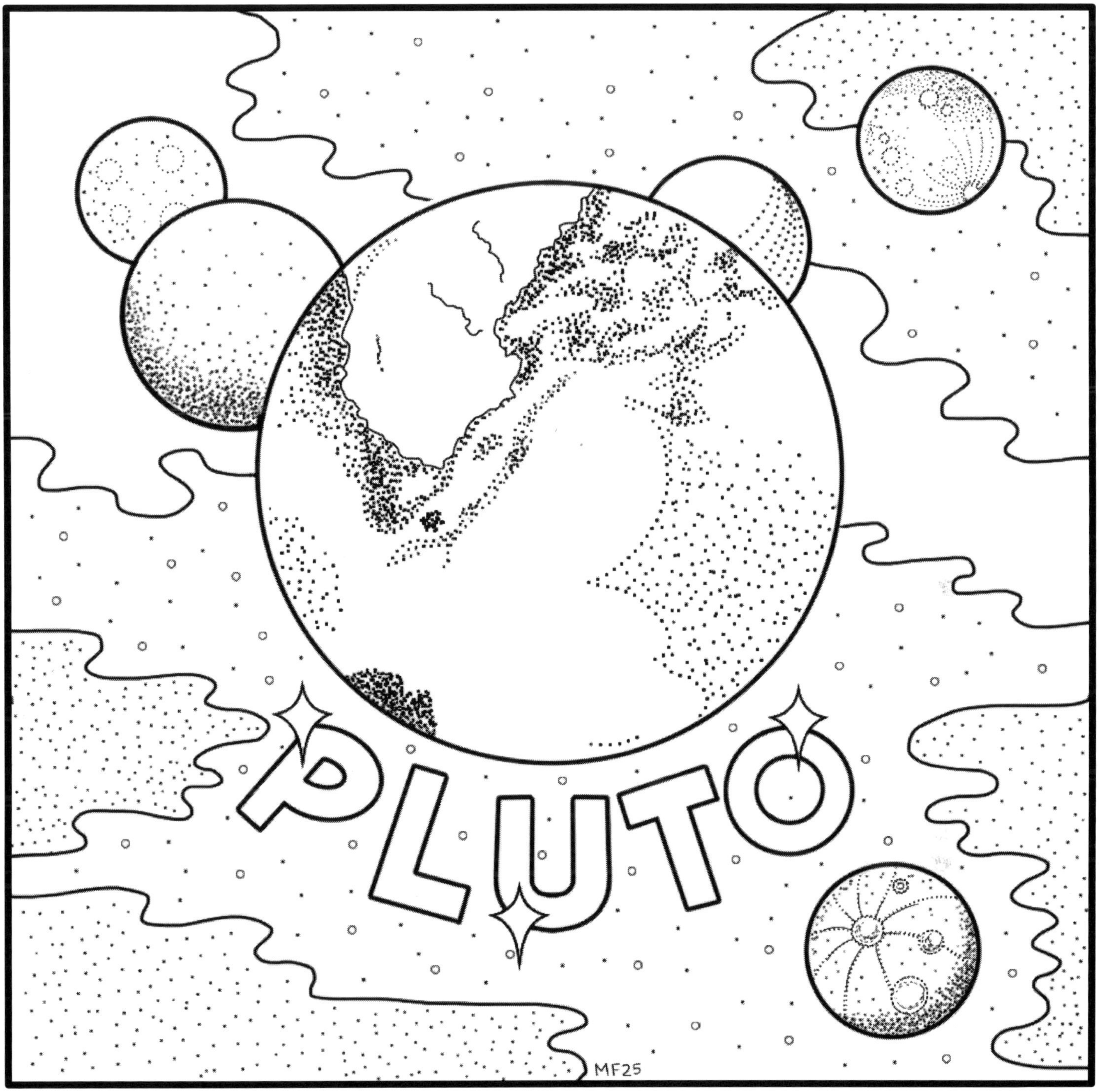

Name Origin:	Diameter:	State of Matter:
Interesting Facts:	Gravity:	Moons:
	Rings:	Km from sun:

DWARF PLANET

A dwarf planet is a celestial body that orbits the Sun and is large enough for gravity to make it nearly round, but unlike a planet, it has not cleared other objects from its orbital path. Dwarf planets are not moons and are typically found in regions such as the asteroid belt or Kuiper Belt. They help scientists understand the formation and evolution of the solar system.

Ceres

Ceres is the largest object in the asteroid belt between Mars and Jupiter, with a diameter of about 940 kilometers, and it orbits the Sun at an average distance of approximately 414 million kilometers. It is made of rock and ice and may contain subsurface water. Ceres has no known moons or rings, and its surface gravity is about 0.27 m/s^2. It was discovered in 1801 by astronomer Giuseppe Piazzi and named after Ceres, the Roman goddess of agriculture and grain, reflecting its location among early "planet-like" objects.

Eris

Eris is a distant dwarf planet in the scattered disk, with a diameter of about 2,326 kilometers, and it orbits the Sun at an average distance of roughly 10 billion kilometers. It is made of ice and rock, has one known moon, no rings, and surface gravity of about 0.82 m/s^2. Eris was discovered in 2005 and named after Eris, the Greek goddess of chaos and discord, reflecting the scientific debate it sparked that led to the redefinition of planets.

Haumea

Haumea is a fast-spinning dwarf planet in the Kuiper Belt with an unusual elongated shape and an average diameter of about 1,632 kilometers. It orbits the Sun at approximately 6.5 billion kilometers, is made of rock and ice, and has two moons, a thin ring, and surface gravity of about 0.44 m/s^2. Haumea was discovered in 2004 and named after Haumea, the Hawaiian goddess of fertility and childbirth, honoring both its rapid rotation and the Hawaiian location of the discovery telescope.

Makemake

Makemake is a dwarf planet in the Kuiper Belt with a diameter of about 1,430 kilometers, orbiting the Sun at an average distance of roughly 6.8 billion kilometers. It is composed of ice and rock, has one known moon, no rings, and surface gravity of about 0.5 m/s^2. Makemake was discovered in 2005 and named after Makemake, the creator god of the Rapa Nui people of Easter Island, reflecting the timing of its discovery near Easter.

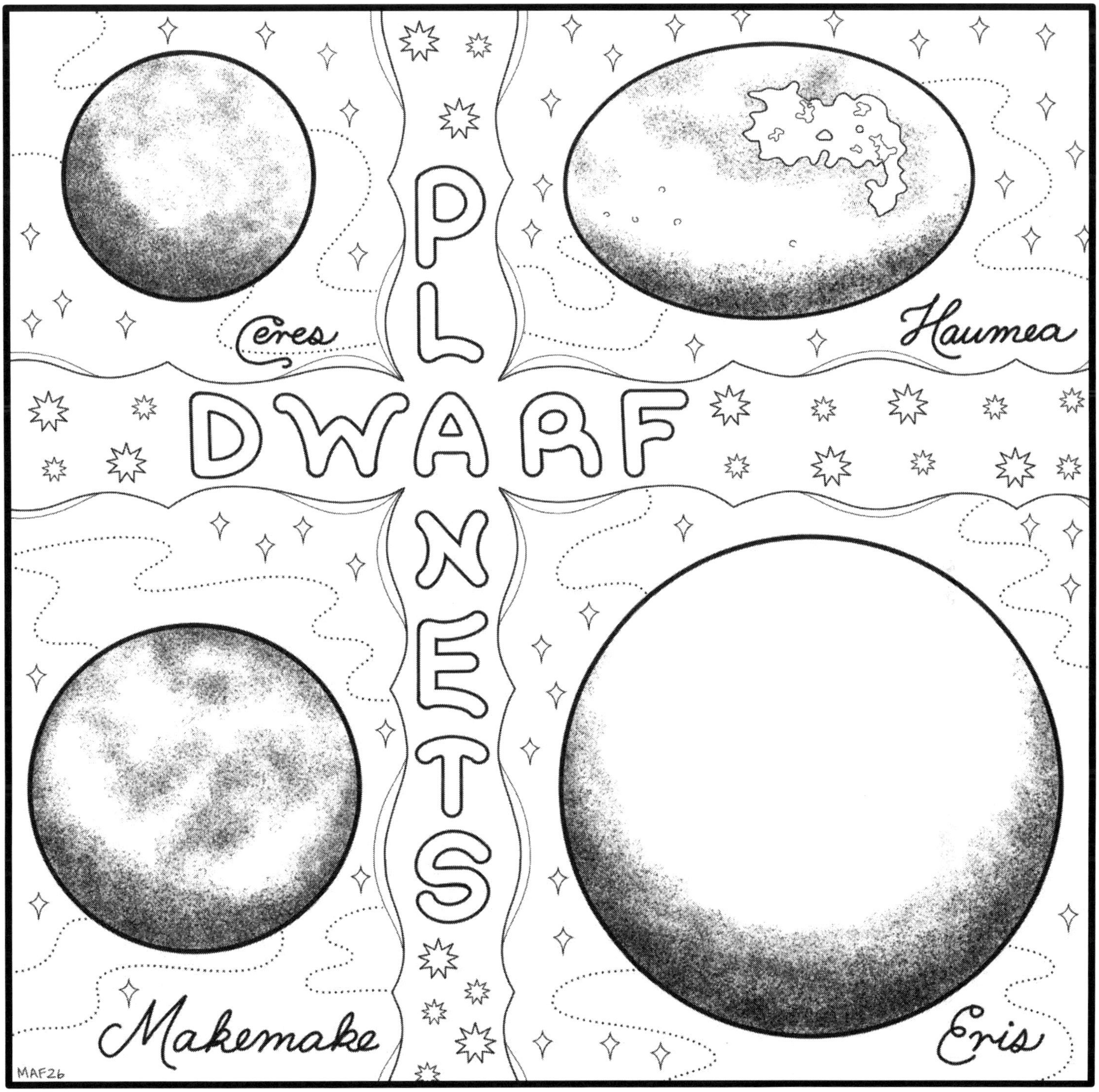

What characteristics must a celestial body have in order to be classified as a dwarf planet, and how does this differ from a planet?	How does the location of each dwarf planet (Ceres, Eris, Haumea, and Makemake) relate to its region of the solar system?
Choose one dwarf planet from the reading and describe its physical features, including composition, moons or rings, and surface gravity.	**Why was the discovery of Eris significant to the scientific understanding and classification of planets?**

KUIPER BELT

The Kuiper Belt is a vast region of the outer solar system located beyond the orbit of Neptune, extending from about 4.5 to 7.5 billion kilometers from the Sun (roughly 30 to 50 astronomical units). Unlike the asteroid belt, the Kuiper Belt is much larger and thicker, forming a wide, disk-shaped region that extends above and below the plane of the planets. Even so, it is still very spread out, and objects within it are separated by enormous distances.

The Kuiper Belt is made primarily of icy bodies composed of frozen water, methane, ammonia, and rock. These objects range in size from small icy fragments to large dwarf planets such as Pluto, Haumea, Makemake, and Eris. Many Kuiper Belt objects have irregular shapes and highly elliptical or tilted orbits, reflecting the complex gravitational history of the outer solar system.

Scientists believe the Kuiper Belt formed about 4.6 billion years ago as leftover material from the formation of the planets. While the inner planets and gas giants formed closer to the Sun, icy material remained in the colder outer regions. Neptune's gravity played a major role in shaping the Kuiper Belt, scattering objects into their current orbits and preventing them from forming a single large planet.

The Kuiper Belt was theorized in the 1950s but was not confirmed until 1992, when astronomers David Jewitt and Jane Luu discovered the first Kuiper Belt object beyond Pluto. Since then, thousands of Kuiper Belt objects have been identified, dramatically expanding scientific understanding of the outer solar system.

One of the most important features of the Kuiper Belt is that it is the source of many short-period comets, which orbit the Sun in less than 200 years. The Kuiper Belt also represents a relatively unchanged record of early solar system material, making it a key region for studying planetary formation and evolution. Far from being empty or inactive, the Kuiper Belt is a dynamic and essential part of our solar system.

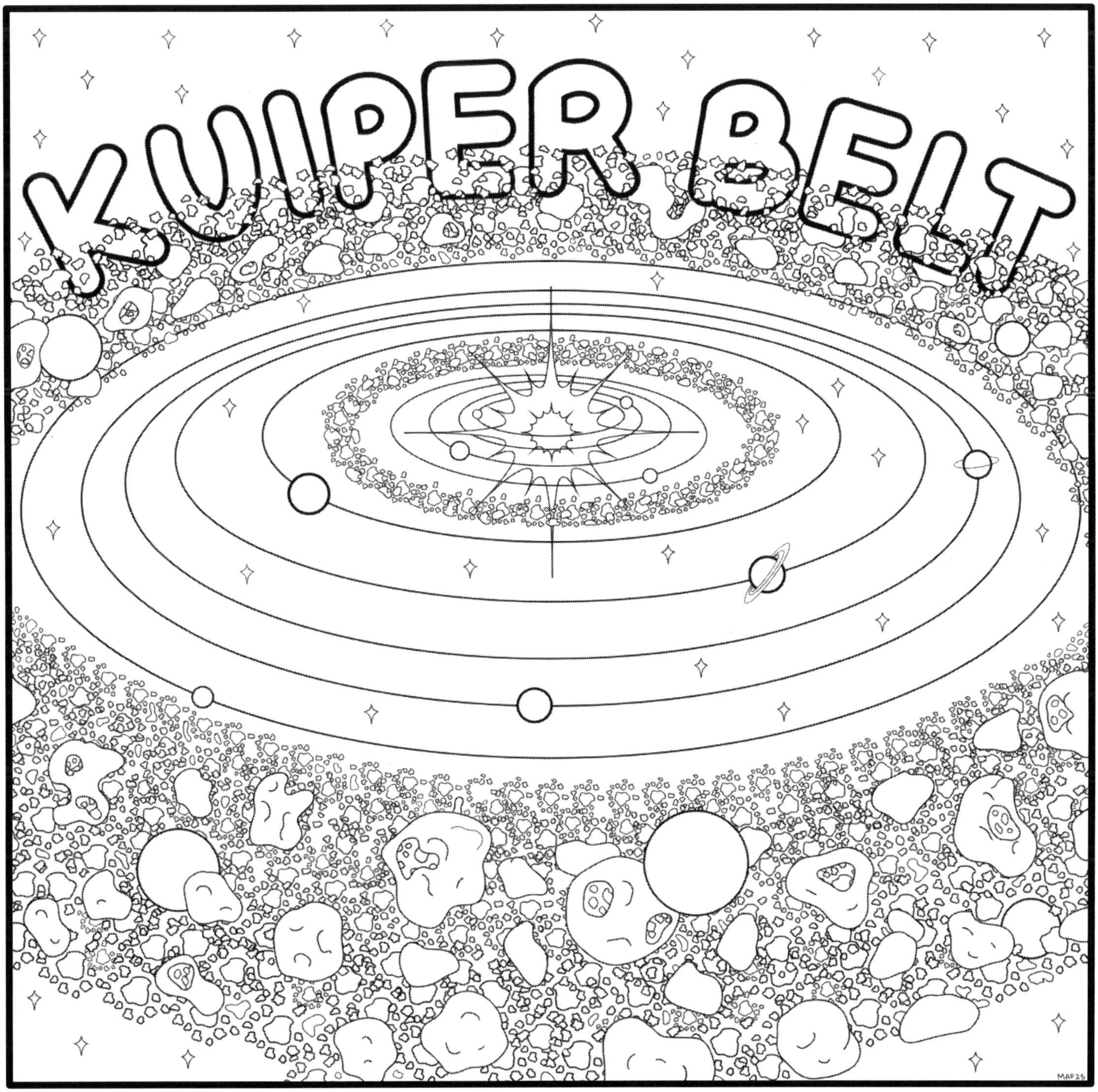

Where is the Kuiper Belt located in the solar system, and how does it compare in size and structure to the asteroid belt?

What types of materials make up most Kuiper Belt objects, and what range of objects can be found there?

How did Neptune's gravity influence the formation and current structure of the Kuiper Belt?

Why is the Kuiper Belt important to scientists studying the early solar system and comets?

OORT CLOUD

The Oort Cloud is a distant and vast region of the solar system that surrounds it like a spherical shell. It lies far beyond the Kuiper Belt, beginning thousands of astronomical units from the Sun and extending up to tens of thousands of astronomical units away. Because of this immense distance, the Oort Cloud is the most remote region of our system and cannot be observed directly with telescopes. Its existence is supported by mathematical models and observations of comet behavior.

The Oort Cloud is made up primarily of icy objects composed of frozen water, methane, ammonia, and other volatile materials mixed with rock. These objects are remnants left over from the formation of the solar system about 4.6 billion years ago. Unlike the disk-shaped Kuiper Belt, the Oort Cloud is thought to be roughly spherical, extending in all directions around the solar system. This shape suggests that gravitational influences from nearby stars and the Milky Way galaxy helped shape its structure.

Scientists believe the Oort Cloud formed when giant planets such as Jupiter and Saturn scattered icy material outward during the early stages of planetary formation. Many of these objects were pushed far enough away that they became weakly bound to the Sun by gravity. Over time, these distant objects settled into the Oort Cloud, where they remain mostly unchanged, preserving clues about the early solar system.

One of the most important roles of the Oort Cloud is that it is believed to be the source of long-period comets. These comets have highly elongated orbits and can take thousands or even millions of years to complete one trip around the Sun. When gravitational disturbances from passing stars or galactic forces occur, some Oort Cloud objects are sent inward toward the Sun, becoming visible comets. Studying these comets helps scientists learn about the composition and history of the solar system.

The Oort Cloud is important because it represents a frozen record of the solar system's earliest materials. By studying the comets that originate from it, astronomers gain insight into how planets formed and how the solar system evolved. Although it remains mysterious and distant, the Oort Cloud plays a crucial role in understanding the outer boundaries of our solar system.

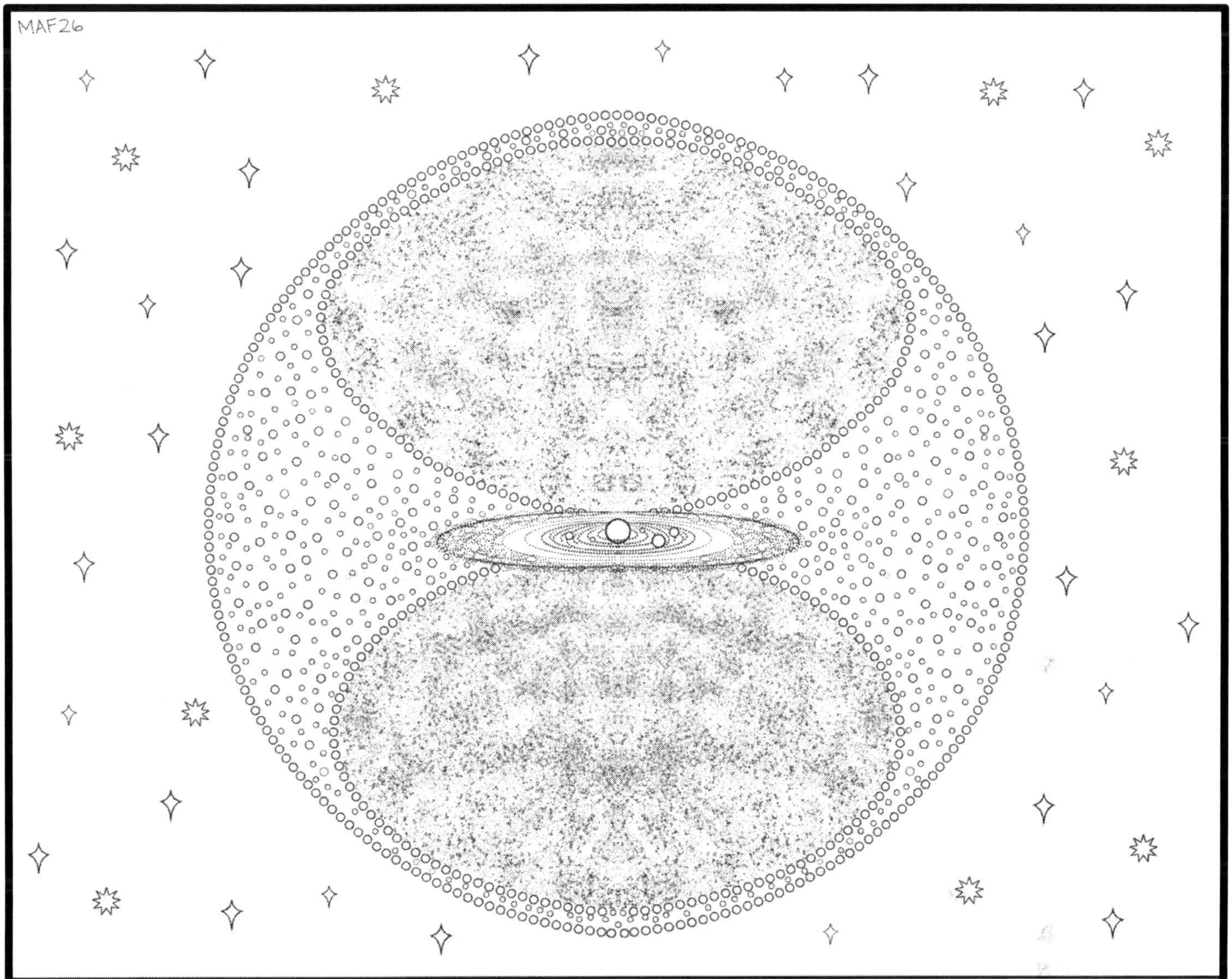

Oort Cloud

Where is the Oort Cloud located, and how does it differ in shape from the Kuiper Belt?	What materials make up most objects in the Oort Cloud?
How did the gas giant planets influence the formation of the Oort Cloud?	Why is the Oort Cloud an important source of scientific information about the early solar system?

EXOPLANET

An exoplanet, also known as an extrasolar planet, is a planet that orbits a star outside of our solar system. While the planets in our solar system orbit the Sun, exoplanets orbit other stars within our galaxy and beyond. Astronomers study exoplanets to better understand how planetary systems form and how common planets may be throughout the universe.

Exoplanets vary widely in size, composition, and orbit. Some are gas giants similar to Jupiter, while others are rocky like Earth. Certain exoplanets orbit very close to their stars, making them extremely hot, while others are located farther away in cold regions. Some exoplanets are found in the habitable zone of their stars, where temperatures may allow liquid water to exist on the planet's surface, a key factor in the search for life.

Astronomers detect exoplanets using several scientific methods. One common method is the transit method, which occurs when a planet passes in front of its star and causes a small decrease in the star's brightness. Another method is the radial velocity method, which detects slight movements of a star caused by the gravitational pull of an orbiting planet. Advanced space telescopes and instruments have greatly increased the number of known exoplanets over time.

Exoplanets are important because they help scientists understand the diversity of planetary systems and how they compare to our own. Studying exoplanets provides insight into planet formation, atmospheric conditions, and the potential for life beyond Earth. As technology improves, astronomers continue to discover new exoplanets, expanding knowledge of the universe and humanity's place within it.

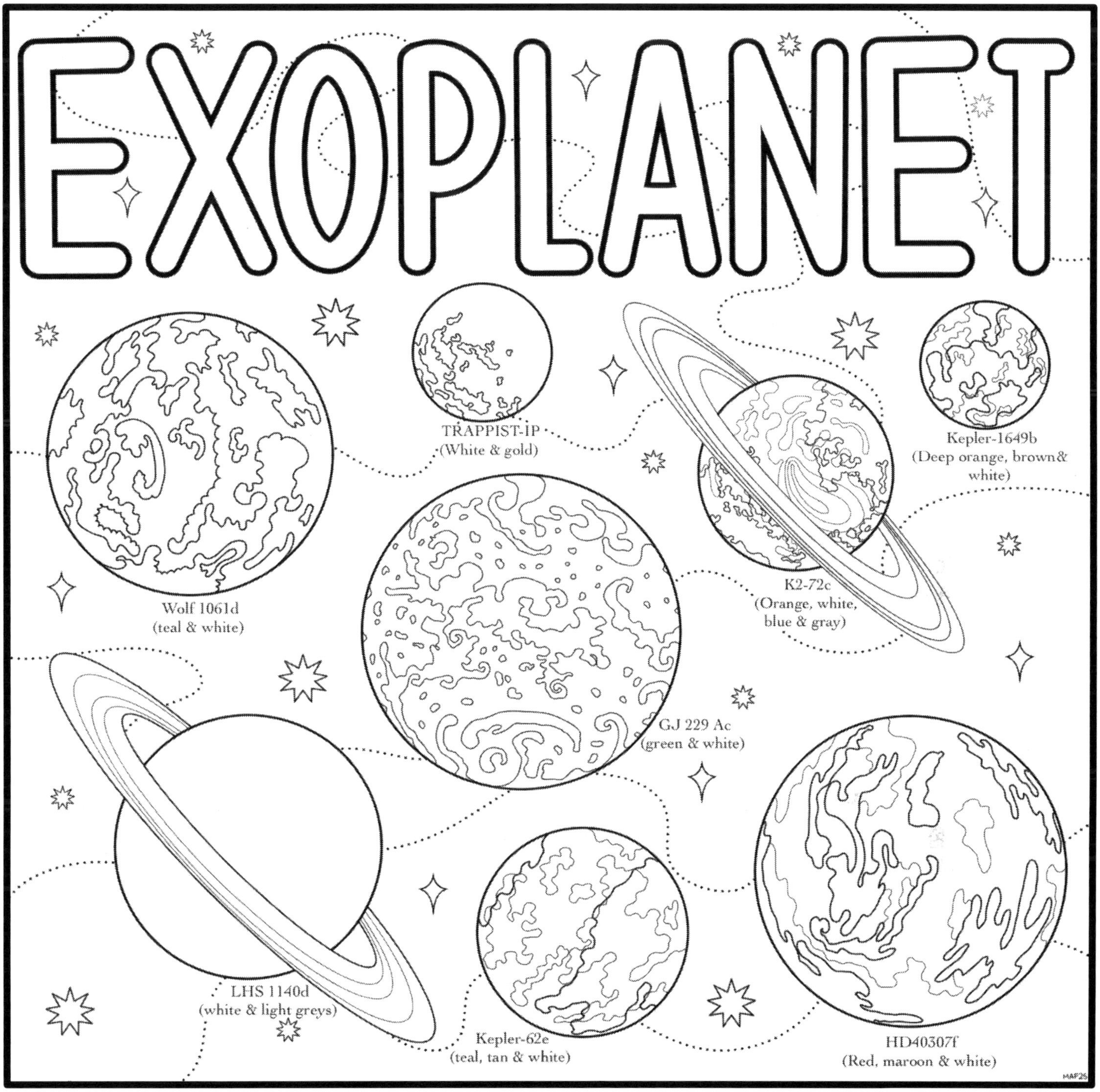

What is an exoplanet, and how is it different from planets in our solar system?	Describe two ways in which exoplanets can differ from one another.
What is one method astronomers use to detect exoplanets, and how does it work?	Why are exoplanets important to the study of astronomy and the search for life?

ORBIT

An orbit is the curved path that one object in space follows as it moves around another object due to gravity. In astronomy, orbits occur when the gravitational pull of a larger object, such as a star or planet, attracts a smaller object while the smaller object continues moving forward. This balance between gravity pulling inward and forward motion carrying the object sideways causes the object to follow a repeating path rather than falling straight into the larger body. Orbits can take different shapes depending on the speed of the object and the strength of gravity. Most orbits are elliptical, meaning they are slightly oval rather than perfectly circular. For example, planets orbit stars in elliptical paths, and moons orbit planets in similar ways. The closer an object is to the body it orbits, the stronger the gravitational pull and the faster it must move to stay in orbit.

Orbits are found throughout the universe and play a key role in how celestial systems function. Earth orbits the Sun, which allows for predictable seasons and a stable climate. The Moon orbits Earth, influencing ocean tides through its gravitational pull. Artificial satellites also orbit Earth and are used for communication, weather monitoring, navigation, and scientific research. Without stable orbits, these systems would not function properly.

Scientists study orbits to understand the motion of objects in space and the forces that control them. By studying orbits, astronomers can calculate the mass of planets and stars, predict eclipses, and send spacecraft to precise locations in space. Understanding orbits is essential for space exploration, satellite technology, and explaining how gravity shapes the structure of the universe.

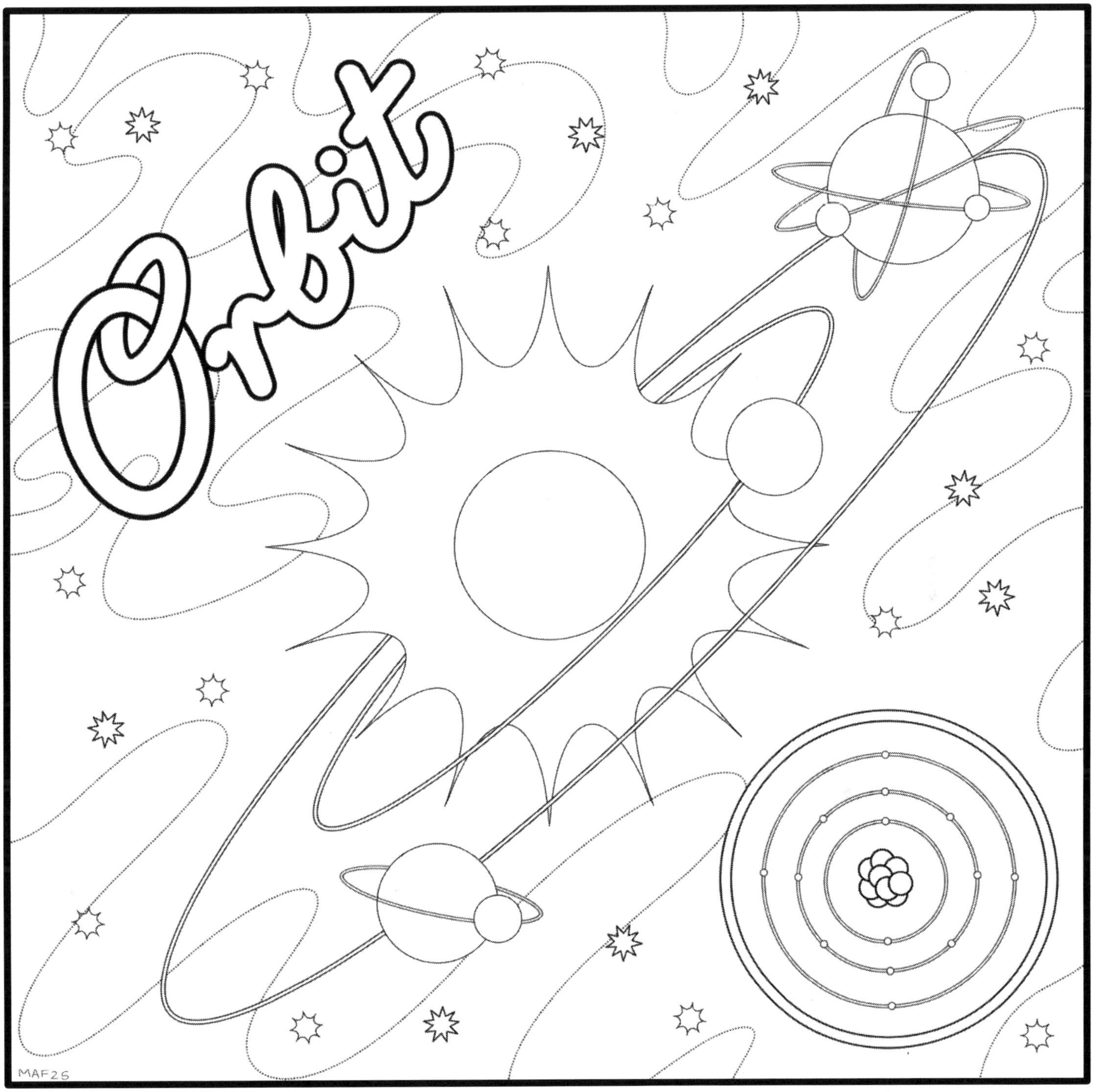

What is an orbit, and what two factors work together to create it?

Why are most orbits elliptical rather than perfectly circular?

How do orbits affect natural systems such as Earth and the Moon?

Why is understanding orbits important for space exploration and satellite use?

STAR

Stars are massive, glowing spheres of hot gas, primarily hydrogen and helium, held together by gravity. They produce energy through nuclear fusion, a process in which hydrogen atoms fuse into helium in the star's core, releasing enormous amounts of light and heat. This energy pushes outward, balancing gravity's inward pull and keeping the star stable for most of its life. Stars vary widely in size, temperature, brightness, color, and lifespan, depending mainly on their mass.

Astronomers classify stars by their surface temperature and brightness, which are closely related to color. Hotter stars appear blue or white, while cooler stars appear yellow, orange, or red. One common classification system groups stars as O, B, A, F, G, K, and M, from hottest to coolest. Most stars in the universe, including the Sun, spend the majority of their lives in a stable phase called the main sequence, where hydrogen fusion occurs steadily in the core.

The Sun is a G-type main-sequence star, often called a yellow dwarf. It is about 4.6 billion years old and provides the energy that drives Earth's climate, weather, and life. Compared to many stars, the Sun is medium-sized and relatively ordinary, yet it appears much brighter to us because it is far closer than any other star.

Stars are extremely numerous. Astronomers estimate that there are about 100–400 billion stars in the Milky Way galaxy alone, and with hundreds of billions of galaxies in the observable universe, the total number of stars is estimated to be on the order of 10^{22} to 10^{24} stars. Because of their vast distances, stars appear as tiny points of light, but each one is a powerful and complex object that plays a role in shaping galaxies and producing the elements needed for planets and life.

What is a star, how is it different from a planet?	Describe a star's life cycle.
List and explain at least three star classifications.	How do scientists determine the temperature and color of a star?

STELLAR LIFE CYCLE

Stars are born, evolve, and eventually die in a process known as the stellar life cycle. The exact path a star follows depends largely on its mass. All stars begin their lives in large clouds of gas and dust called nebulae. When gravity causes parts of a nebula to collapse, the material heats up and forms a protostar. Once the core becomes hot enough for nuclear fusion to begin, a true star is born.

For most of a star's life, it remains in the main sequence phase, where hydrogen is fused into helium in the core. Smaller, low-mass stars burn their fuel slowly and can remain stable for billions to trillions of years. Larger, high-mass stars burn fuel much faster and live much shorter lives, sometimes only millions of years. As hydrogen fuel runs low, the balance between gravity and outward pressure changes, and the star begins to evolve.

Low- to medium-mass stars, such as the Sun, expand into red giants as their outer layers cool and spread outward. Eventually, these stars shed their outer layers into space, forming a planetary nebula, while the core remains as a dense white dwarf. Over extremely long periods of time, the white dwarf cools and fades.

High-mass stars follow a more dramatic path. After becoming red supergiants, they undergo violent explosions called supernovae, releasing vast amounts of energy and scattering heavy elements into space. The remaining core may collapse into a neutron star or, if massive enough, a black hole. These explosive deaths help create new nebulae, allowing the cycle of star formation to continue.

The life cycles of stars are essential to the universe because stars create most of the chemical elements found in planets, oceans, and living organisms. Every atom of carbon, oxygen, and iron on Earth was once formed inside a star, making stellar evolution a key process in the story of the cosmos.

Stellar Life Cycle

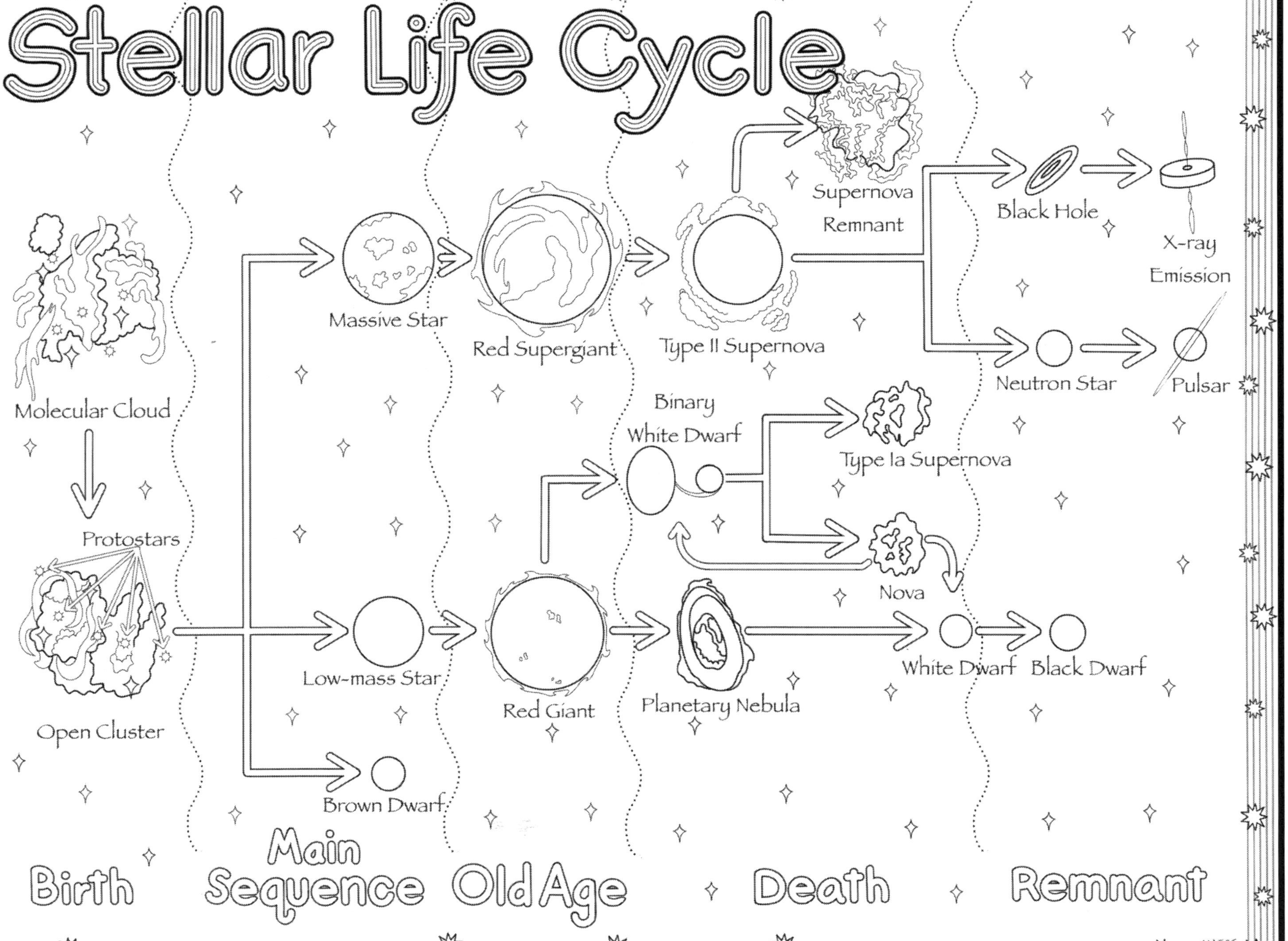

LIGHT YEAR

A light-year is a unit of distance used in astronomy to measure the vast spaces between objects in the universe. Despite its name, a light-year does not measure time. Instead, it represents the distance that light travels in one year. Light moves extremely fast, traveling at a speed of about 300,000 kilometers per second in a vacuum. Because distances in space are so large, astronomers use light-years instead of kilometers or miles to describe how far apart stars and galaxies are.
In one year, light travels approximately 9.46 trillion kilometers. This means that an object one light-year away is so distant that the light we see from it today actually left that object one year ago. When astronomers observe stars or galaxies many light-years away, they are looking into the past. The farther away an object is, the older the light is that is being observed, which allows scientists to study how the universe has changed over time.

Light-years are especially useful for describing distances beyond our solar system. For example, the nearest star system to Earth after the Sun is several light-years away, and our galaxy is tens of thousands of light-years across. Galaxies beyond the Milky Way are millions or even billions of light-years away. Using light-years helps astronomers compare distances on a scale that makes sense for the size of the universe.

Understanding light-years is important because it helps scientists measure and explain the structure of the universe. By knowing how far away objects are, astronomers can determine their size, brightness, and movement. Light-years also help scientists study the expansion of the universe and the life cycles of stars and galaxies. This measurement plays a key role in understanding how the universe formed and continues to evolve.

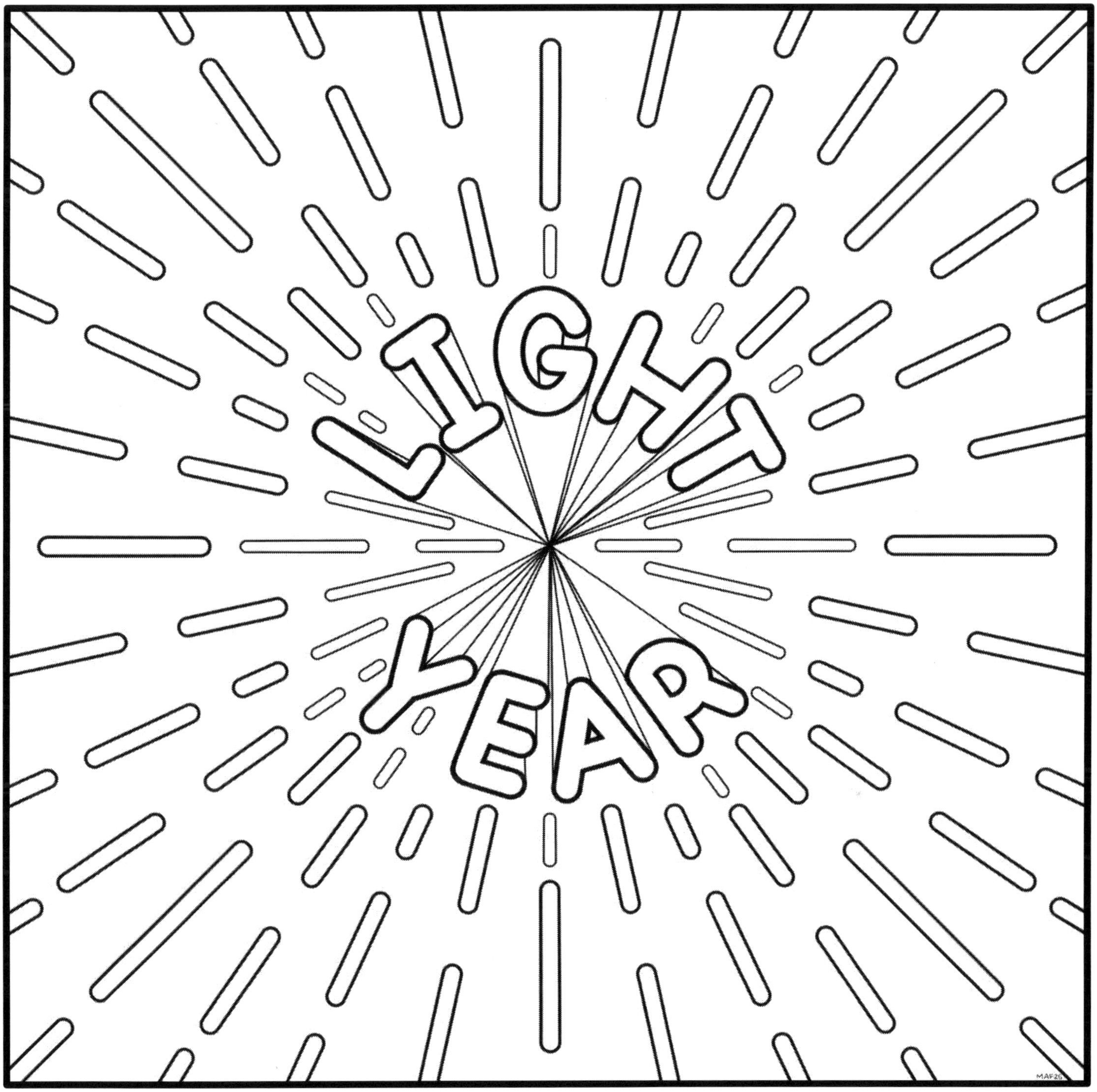

What is a light-year, and what does it measure?	Why is a light-year used instead of kilometers or miles in astronomy?
What does it mean when astronomers say they are looking into the past when observing distant objects?	How do light-years help scientists understand the structure of the universe?

SUPERNOVA

A supernova is a powerful and dramatic explosion that occurs at the end of a star's life cycle. In astronomy, supernovae are important events because they release enormous amounts of energy and matter into space, briefly making a single star shine brighter than an entire galaxy. These explosions mark the death of certain types of stars and play a major role in shaping the structure and chemical makeup of the universe.

Supernovae occur when a star can no longer maintain a balance between gravity and the energy produced by nuclear fusion in its core. During most of a star's life, fusion creates outward pressure that balances the inward pull of gravity. When the fuel in the star's core is exhausted, this balance is disrupted. Gravity causes the core to collapse, triggering a sudden and violent explosion that blasts the star's outer layers into space.

There are two main types of supernovae studied in astronomy. One type occurs in massive stars that are much larger than the Sun. When these stars run out of fuel, their cores collapse rapidly, causing a core-collapse supernova. The other type occurs in a binary star system, where a dense white dwarf star pulls material from a nearby companion star. When the white dwarf gains too much mass, it becomes unstable and explodes in a thermonuclear supernova. Although the causes are different, both types release vast amounts of energy and matter. Supernovae are extremely important to the universe because they create and spread heavy elements. Elements such as iron, gold, and uranium are formed during these explosions and are scattered into space when the star explodes. Over time, this material becomes part of new stars, planets, and even living organisms. Without supernovae, many of the elements necessary for planets and life would not exist.

In addition to producing elements, supernovae influence the formation of stars and galaxies. The shock waves from an explosion can compress nearby clouds of gas and dust, triggering the formation of new stars. Supernovae also leave behind dense remnants such as neutron stars or black holes, which continue to affect their surroundings through intense gravity and radiation.

Astronomers study supernovae using telescopes that detect visible light, X-rays, and other forms of electromagnetic radiation. By observing supernovae, scientists learn about stellar life cycles, the expansion of the universe, and the forces that shape galaxies. These powerful explosions provide key evidence for understanding how matter and energy move through space over time.

A supernova represents both an ending and a beginning in astronomy. While it marks the death of a star, it also contributes to the creation of new stars, planets, and the elements that make up the universe. Through the study of supernovae, scientists gain insight into the dynamic and ever-changing nature of space.

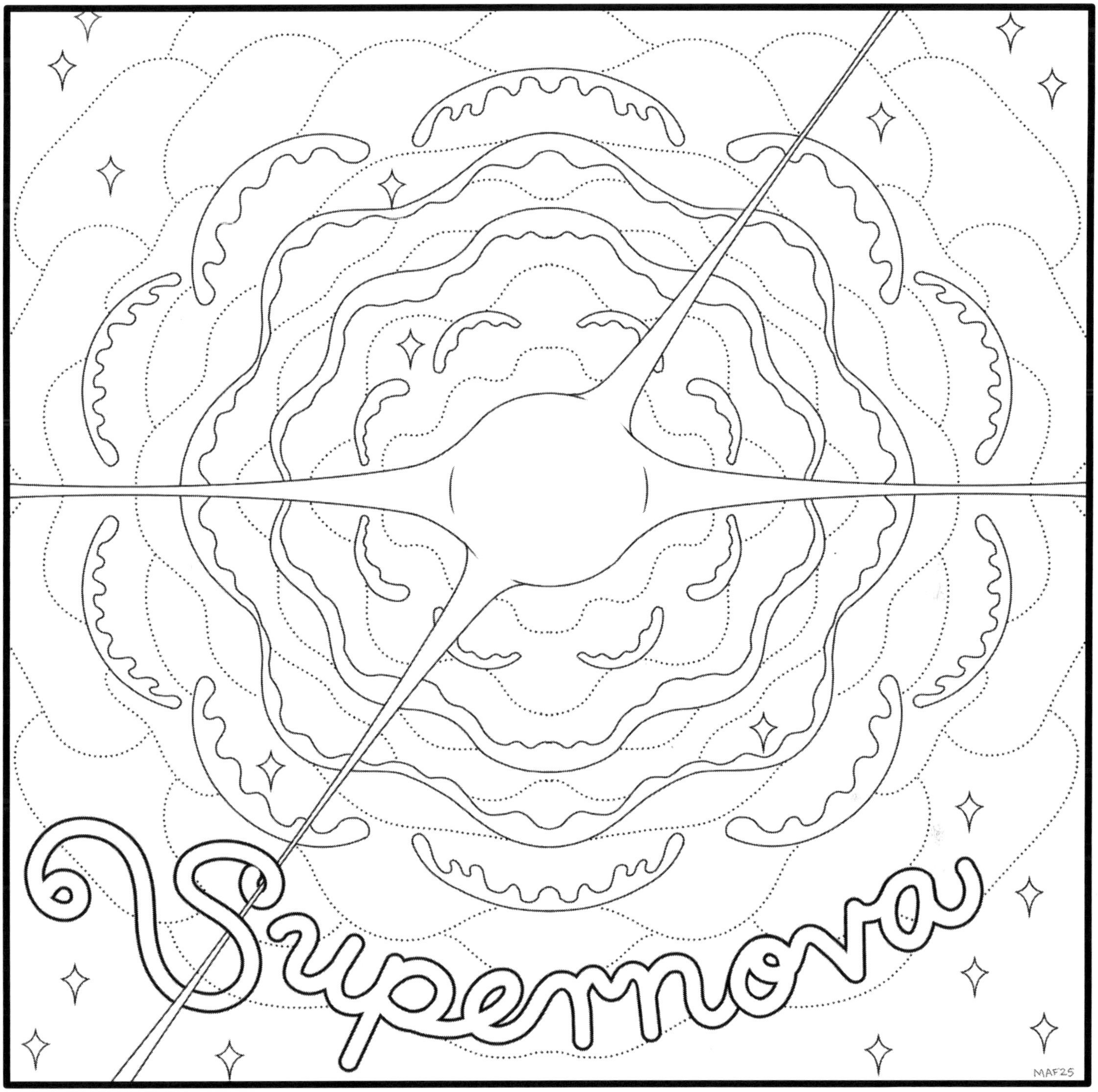

What causes a supernova to occur in the life cycle of a star?	How are the two main types of supernovae different from each other?
Why are supernovae important for the formation of elements in the universe?	How do supernova explosions influence the formation of new stars and galaxies?

CONSTELLATION

A constellation is a recognized pattern of stars in the night sky that forms an imaginary shape or outline when viewed from Earth. In astronomy, constellations are used to organize and map the sky. The stars within a constellation may appear close together, but they are often very far apart in space and are not physically connected. Their patterns are the result of perspective from Earth.

Historically, many constellations were named after animals, objects, or mythological characters by ancient civilizations. These early skywatchers used constellations to track the seasons, navigate during travel, and tell stories about the night sky. While these cultural interpretations vary, modern astronomy uses constellations primarily as reference regions rather than pictures.

In science, the sky is divided into 88 officially recognized constellations. Each constellation represents a specific area of the sky, not just a group of stars. Astronomers use these defined regions to locate stars, planets, galaxies, and other celestial objects. When a star or planet is said to be “in” a constellation, it means it appears within that region of the sky as seen from Earth.

Constellations are important because they help astronomers communicate locations in space and study the motion of celestial objects. As Earth orbits the Sun, different constellations become visible at different times of the year. This predictable pattern allows astronomers to observe changes in the sky and helps students and scientists learn how Earth’s movement affects what we see in space.

Although constellations are not physical structures, they play an essential role in astronomy by providing a framework for understanding and exploring the night sky. They connect observational science with Earth’s position in space and continue to be a valuable tool for studying the universe.

Constellation

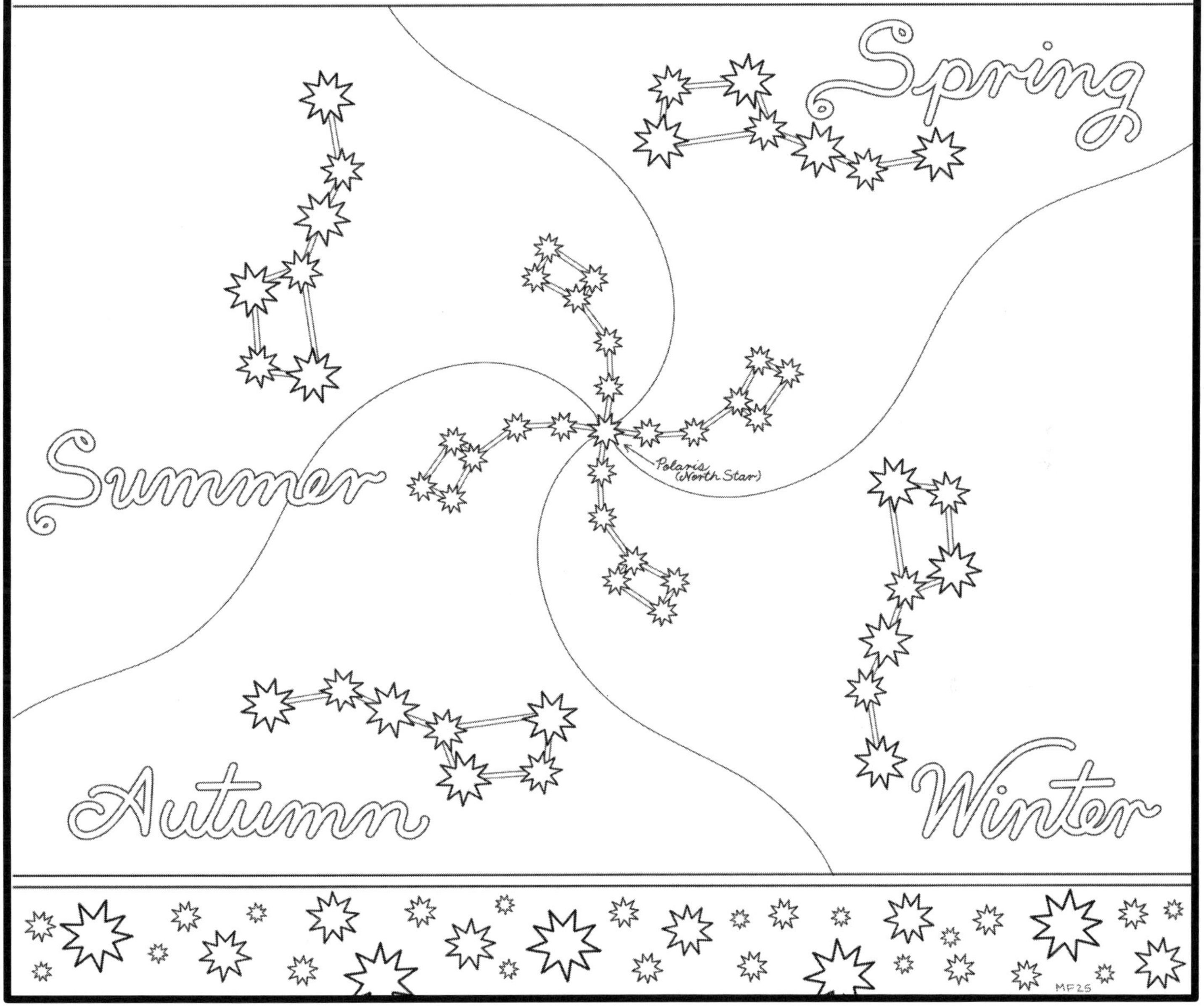

What is a constellation, and why do its stars only appear close together from Earth?	How were constellations used by ancient civilizations?
How do modern astronomers use constellations today?	Why do different constellations appear in the night sky during different seasons?

COMET

Comets are small icy bodies that orbit the Sun and are often described as "dirty snowballs" made of frozen gases, rock, and dust. When a comet travels close to the Sun, heat causes its ice to vaporize, forming a glowing cloud called a coma and one or more tails that can stretch millions of kilometers through space. These tails always point away from the Sun due to solar radiation and the solar wind.

Comets have been observed and recorded by humans for thousands of years. Ancient civilizations, including the Chinese, Babylonians, and Europeans, carefully documented comet appearances, often interpreting them as omens or signs of major events. The first comet recognized as a recurring object was Halley's Comet, whose periodic nature was identified in 1705 by astronomer Edmond Halley. By studying historical records, Halley correctly predicted that the comet would return in 1758, proving that comets orbit the Sun rather than appearing randomly in the sky.

Comets follow highly elliptical orbits, meaning their paths around the Sun are stretched into long ovals. Based on their orbital periods, comets are classified into two main groups. Short-period comets, having orbits lasting less than 200 years and originate primarily from the Kuiper Belt, a region beyond Neptune. Long-period comets have orbits that take hundreds to millions of years and are believed to originate in the distant Oort Cloud, a spherical region surrounding the solar system. Some comets may pass through the inner solar system only once before being ejected into deep space.

Comets are composed of frozen water, carbon dioxide, carbon monoxide, methane, ammonia, and organic compounds mixed with rocky material and dust. As a comet warms, these substances sublimate directly from solid to gas, carrying dust outward to form the coma and tail. Because comets formed in the cold outer regions of the early solar system, their composition has changed very little over time, making them valuable records of early solar system material.

Scientists study comets using a combination of telescopic observations, space missions, and spectroscopy. Telescopes on Earth and in space track comet motion and brightness, while spectrometers analyze the light from a comet to identify its chemical makeup. Space missions such as Rosetta have flown alongside comets and even landed probes on their surfaces, providing direct measurements of composition, structure, and activity. Studying comet debris that falls to Earth as micrometeorites also helps scientists learn about comet material.

Comets are important because they may have played a role in delivering water and organic molecules to early Earth, possibly contributing to the conditions necessary for life. Their predictable orbits allow astronomers to study gravitational interactions within the solar system, and their ancient composition offers insight into how planets formed. Far from being mysterious omens, comets are now understood as dynamic, informative remnants of the solar system's primitive history.

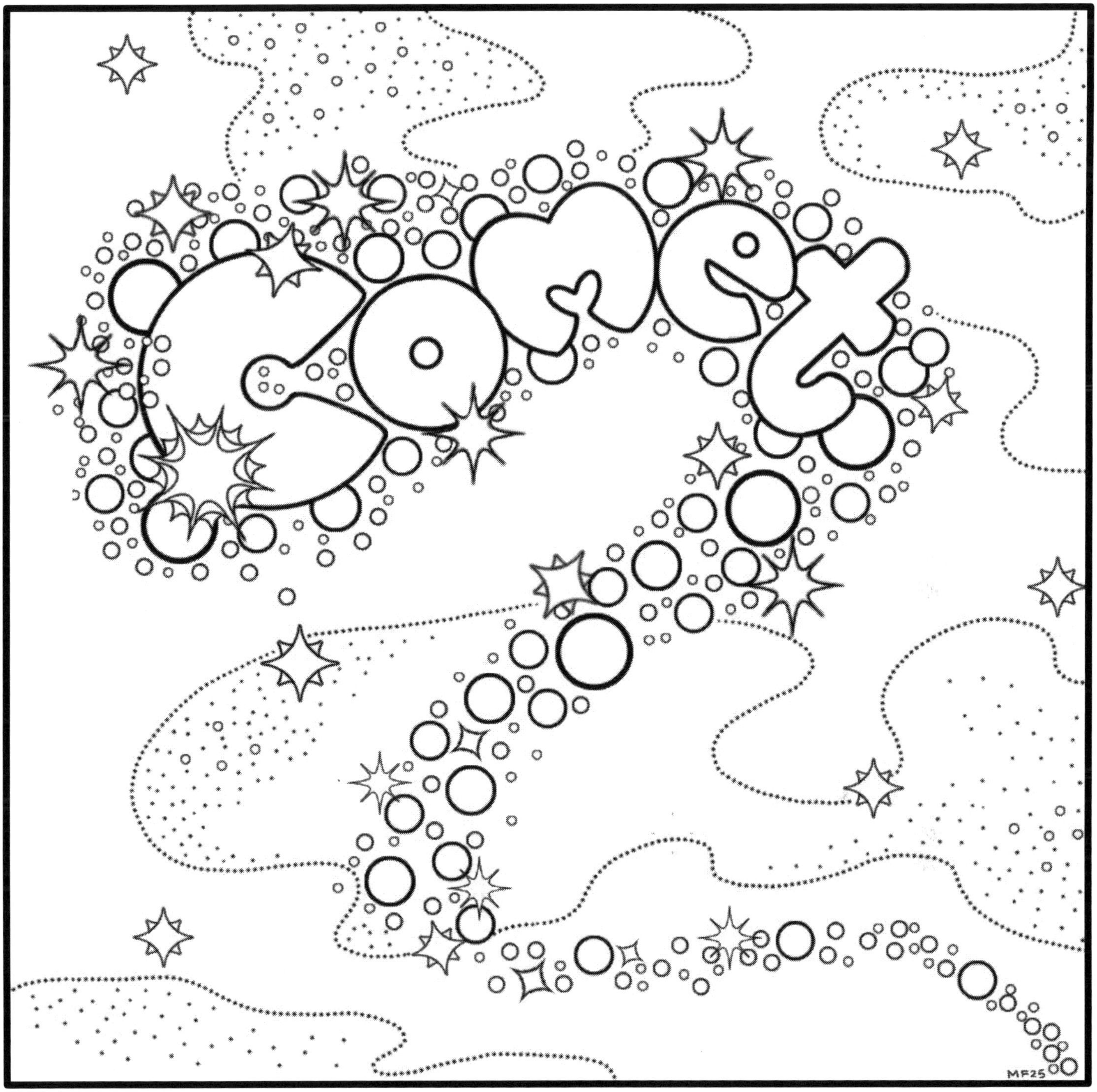

What is a comet made of?	Where do comets come from in the solar system?

What causes comets to develop a tail?	How are comets different from asteroids?

BLACK HOLE

Black holes are regions of space where gravity is so strong that nothing, not even light, can escape. They form when massive amounts of matter collapse into an extremely small area, warping space and time as described by Einstein's theory of general relativity. Although black holes themselves cannot be seen directly, scientists study them by observing their powerful effects on nearby matter, light, and spacetime.

The concept of black holes dates back to the 18th century, when scientists first proposed the idea of objects with gravity strong enough to trap light. Modern understanding developed in the 20th century through advances in physics and astronomy. The first widely accepted black hole candidate, Cygnus X-1, was identified in the 1960s through X-ray observations, providing strong evidence that black holes exist. Astronomers now believe that supermassive black holes lie at the centers of most galaxies, including the Milky Way.

Black holes are classified by mass into stellar-mass black holes, intermediate-mass black holes, and supermassive black holes. Objects such as stars and gas clouds can orbit black holes for millions of years, especially around supermassive ones. Matter drawn toward a black hole forms a spinning accretion disk, where friction heats the material and causes it to emit intense radiation.

The event horizon is the invisible boundary surrounding a black hole beyond which nothing can escape, not even light. It marks the point where the black hole's gravitational pull becomes so strong that the escape velocity exceeds the speed of light. Once matter, light, or radiation crosses the event horizon, it cannot return or send information back to the outside universe.
The event horizon is not a solid surface; it is simply a boundary in space defined by gravity. For a non-rotating black hole, its size is determined by the Schwarzschild radius, which depends on the mass of the black hole, the more massive the black hole, the larger its event horizon.

Because black holes emit no light, scientists detect and study them using indirect evidence and specialized instruments. Space-based X-ray telescopes observe radiation from accretion disks, while optical and infrared telescopes track the motion of stars orbiting invisible massive objects. The Event Horizon Telescope, a global network of radio telescopes, produced the first image of a black hole's shadow in 2019 and later imaged the black hole at the center of the Milky Way. Gravitational wave detectors such as LIGO identify ripples in spacetime caused by black hole collisions, offering direct confirmation of their existence.

Black holes do not have a traditional chemical composition. Instead, they are defined by mass, spin, and electric charge, and any matter that falls in contributes to their mass. Recent discoveries have shown that black holes strongly influence their surroundings. In one notable example, pressure waves detected in hot gas around a supermassive black hole were converted into audible frequencies, sometimes referred to as the "sound of a black hole." These waves demonstrate how black holes transfer energy across vast regions of space. Black holes are no longer just theoretical objects. They are central to understanding stellar evolution, galaxy formation, and the structure of the universe, making them one of the most important topics in modern astronomy.

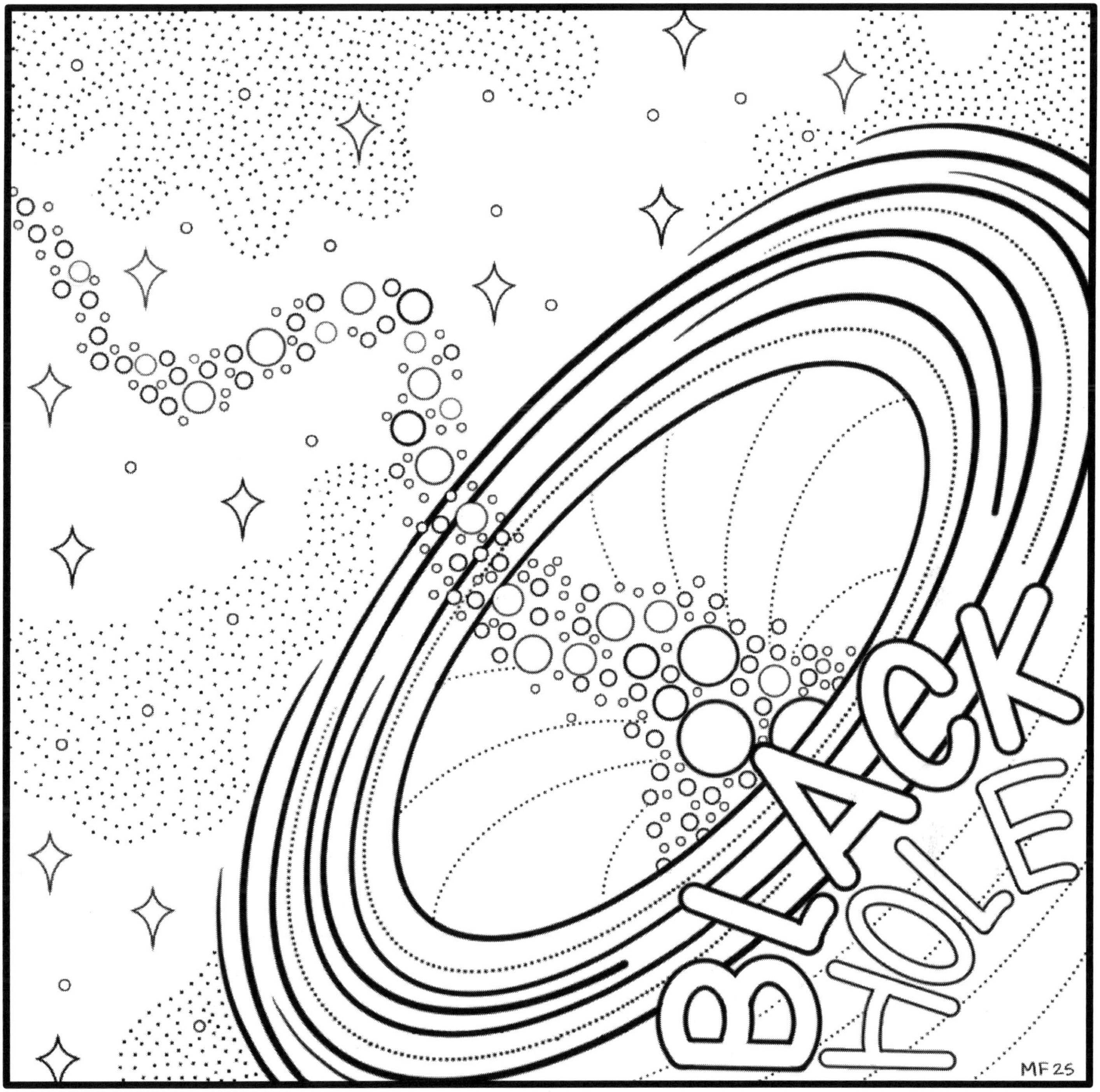

What is a black hole, and how does it form?	How do scientists detect black holes if they cannot be seen directly?
What is the difference between a stellar black hole and a supermassive black hole?	What is the event horizon of a black hole?

ASTEROID

Asteroids are small rocky bodies that orbit the Sun and are remnants from the early formation of the solar system. Often described as leftover building material, asteroids formed about 4.6 billion years ago when dust and rock in the early solar nebula failed to combine into full-sized planets. Many asteroids are found in the asteroid belt between Mars and Jupiter, though others also travel through other regions of the solar system.

The first asteroid ever discovered was Ceres, identified in 1801 by Italian astronomer Giuseppe Piazzi. At the time, Ceres was initially classified as a planet because of its size and orbit. As more similar objects were discovered in the same region of space, scientists realized they were part of a larger population, leading to the creation of the term asteroid, meaning "star-like," because these objects appeared as points of light through early telescopes.

Asteroids follow elliptical orbits around the Sun, though most have relatively low eccentricity and remain in stable paths within the asteroid belt. Orbital periods vary depending on distance from the Sun. Most main-belt asteroids take between 3 and 6 years to complete one orbit. Some asteroids are classified as near-Earth asteroids (NEAs), whose orbits bring them close to Earth, while others follow more elongated or tilted paths due to gravitational interactions with Jupiter or other planets.

Asteroids vary widely in chemical composition, and scientists classify them into major types based on reflectivity and composition. C-type asteroids are rich in carbon and are the most common, S-type asteroids are composed mainly of silicate rock and metal, and M-type asteroids are rich in iron and nickel. Because asteroids have experienced little geological change, their composition closely reflects the material present during the solar system's formation.

Astronomers study asteroids by observing their motion, brightness, and reflected light using telescopes. Changes in brightness reveal an asteroid's rotation rate and shape, while spectral analysis reveals surface composition. Some asteroids occasionally collide with Earth's atmosphere as meteorites, allowing scientists to study asteroid material directly in laboratories.

Modern asteroid research uses ground-based telescopes, space telescopes, and robotic spacecraft missions. Space missions such as OSIRIS-REx and Hayabusa2 have visited asteroids, collected samples, and returned them to Earth for detailed analysis. Radar observations help determine asteroid size, shape, and motion, especially for near-Earth asteroids. These techniques allow scientists to better understand asteroid origins, structure, and potential impact risks.

Asteroids are important because they provide direct evidence of the early solar system and help scientists study how planets formed. Monitoring asteroid orbits also plays a crucial role in planetary defense, helping identify objects that could pose a future threat to Earth. Far from being simple rocks, asteroids are complex, diverse objects that offer valuable insight into the history and evolution of our solar system.

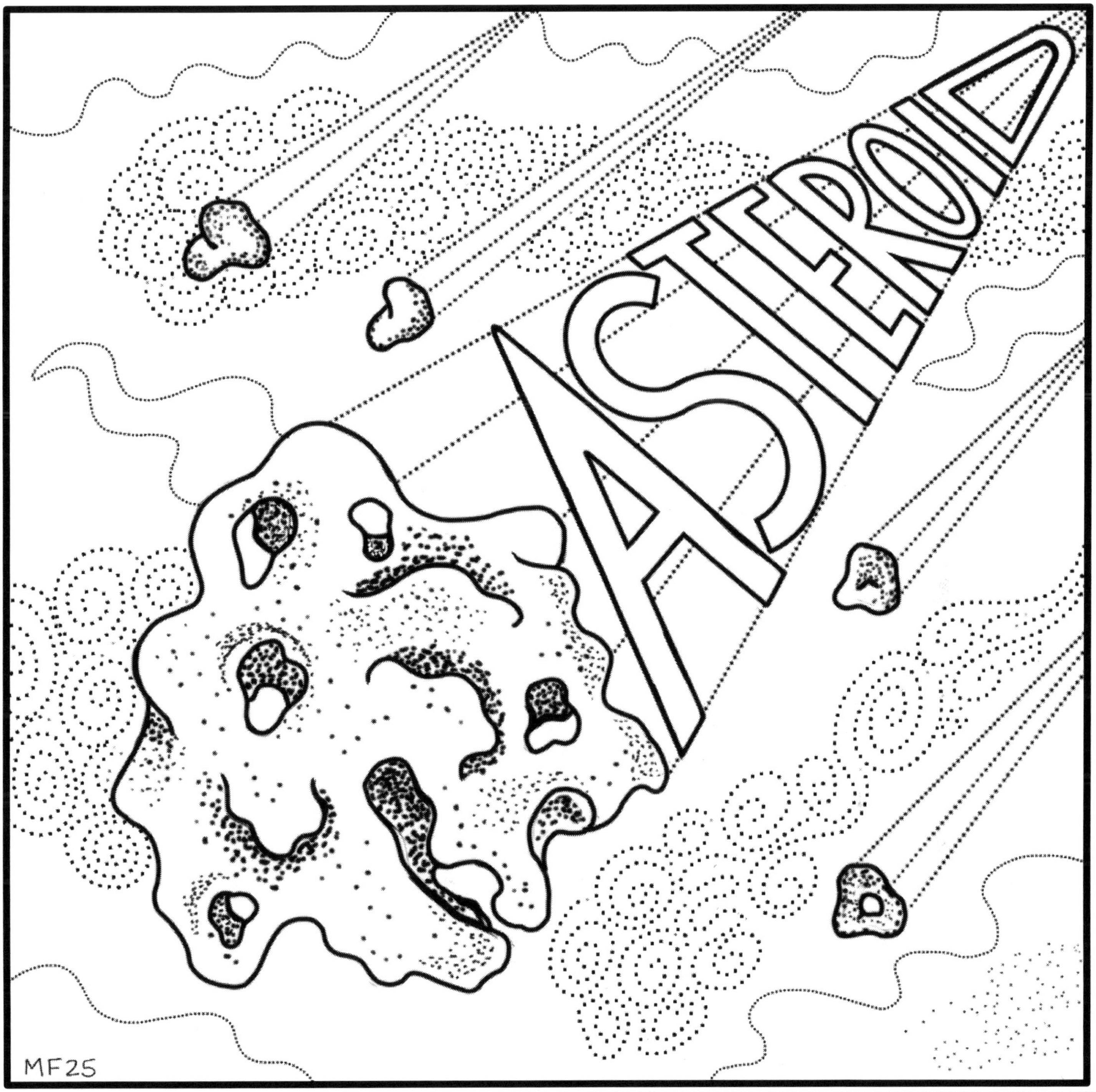

What is an asteroid, what is it composed of?	Where is the asteroid belt located in our solar system?
What are the different types of asteroids?	What are asteroids called when they pass through Earth's atmosphere?

METEOR

Have you ever looked up at the dark night sky and seen a bright green streak of light? Some people call those shooting stars, but they are actually a meteoroid, which is a small rocky or metallic object moving through space, usually originating from asteroids or comets. When a meteoroid enters Earth's atmosphere at high speed, friction with the air causes it to glow, producing a bright streak of light called a meteor, often known as a "shooting star." If part of the object survives its journey through the atmosphere and lands on Earth's surface, it is called a meteorite. These objects provide some of the most direct physical evidence scientists have of material from beyond Earth.

Meteors have been observed and recorded by humans for thousands of years, with descriptions found in ancient Chinese, Greek, and Middle Eastern texts. For much of history, scientists believed meteors were atmospheric phenomena rather than objects from space. This view changed in the late 18th and early 19th centuries, when scientists began collecting and analyzing meteorites. One of the first widely accepted scientific studies of meteorites was conducted by Ernst Chladni in 1794, who argued that rocks falling from the sky were extraterrestrial in origin. His work laid the foundation for modern meteoritics.

Before entering Earth's atmosphere, meteoroids follow elliptical orbits around the Sun, similar to asteroids and comets. Some meteoroids originate from the asteroid belt, while others come from comet debris trails. Meteoroids associated with comets often produce meteor showers, such as the Perseids or Leonids, which occur when Earth passes through streams of debris left behind by comets. Orbital periods can range from a few years to hundreds of years, depending on their source. Meteor showers occur when Earth passes through a trail of dust and small rocky debris left behind by a comet or, in some cases, an asteroid. As these tiny particles enter Earth's atmosphere at high speed, friction with the air causes them to heat up and glow, creating bright streaks of light commonly called "shooting stars." During a meteor shower, many meteors appear to radiate from the same point in the sky, known as the radiant, because they are traveling along parallel paths through space.

Meteorites are classified into three main types based on composition: Stony meteorites, made mostly of silicate minerals Iron meteorites, composed mainly of iron and nickel Stony-iron meteorites, which contain a mixture of rock and metal These materials formed under extreme conditions early in the solar system's history and have remained largely unchanged for billions of years. Scientists study meteors using optical cameras, radar systems, and spectroscopy to analyze their speed, trajectory, and composition as they burn through the atmosphere. Once meteorites are recovered on Earth, they are examined using laboratory techniques such as electron microscopes, mass spectrometers, and radiometric dating. These analyses reveal information about the age, structure, and chemical makeup of the solar system.

Meteorites are extremely valuable scientifically because they are natural samples of early solar system material. Some contain rare metals such as nickel, cobalt, iridium, platinum, and even microscopic diamonds formed under intense pressure. Certain meteorites also contain amino acids and organic compounds, offering clues about the origins of life. Meteorites can also be valuable in monetary terms. Common stony meteorites may sell for $1–5 per gram, while iron meteorites often range from $0.50–$2 per gram. Rare or visually striking meteorites, such as pallasites (a stony-iron type with crystal olivine), can sell for $50–$200 per gram or more. Exceptionally rare meteorites from the Moon or Mars can sell for thousands of dollars per gram, making them some of the most valuable rocks on Earth.

Meteors and meteorites connect Earth directly to the rest of the solar system. They help scientists understand planet formation, asteroid collisions, and the chemical building blocks of life. Whether seen as a brief flash of green light in the sky or studied as a rare rock in a laboratory, meteors provide a powerful link between Earth and the wider universe.

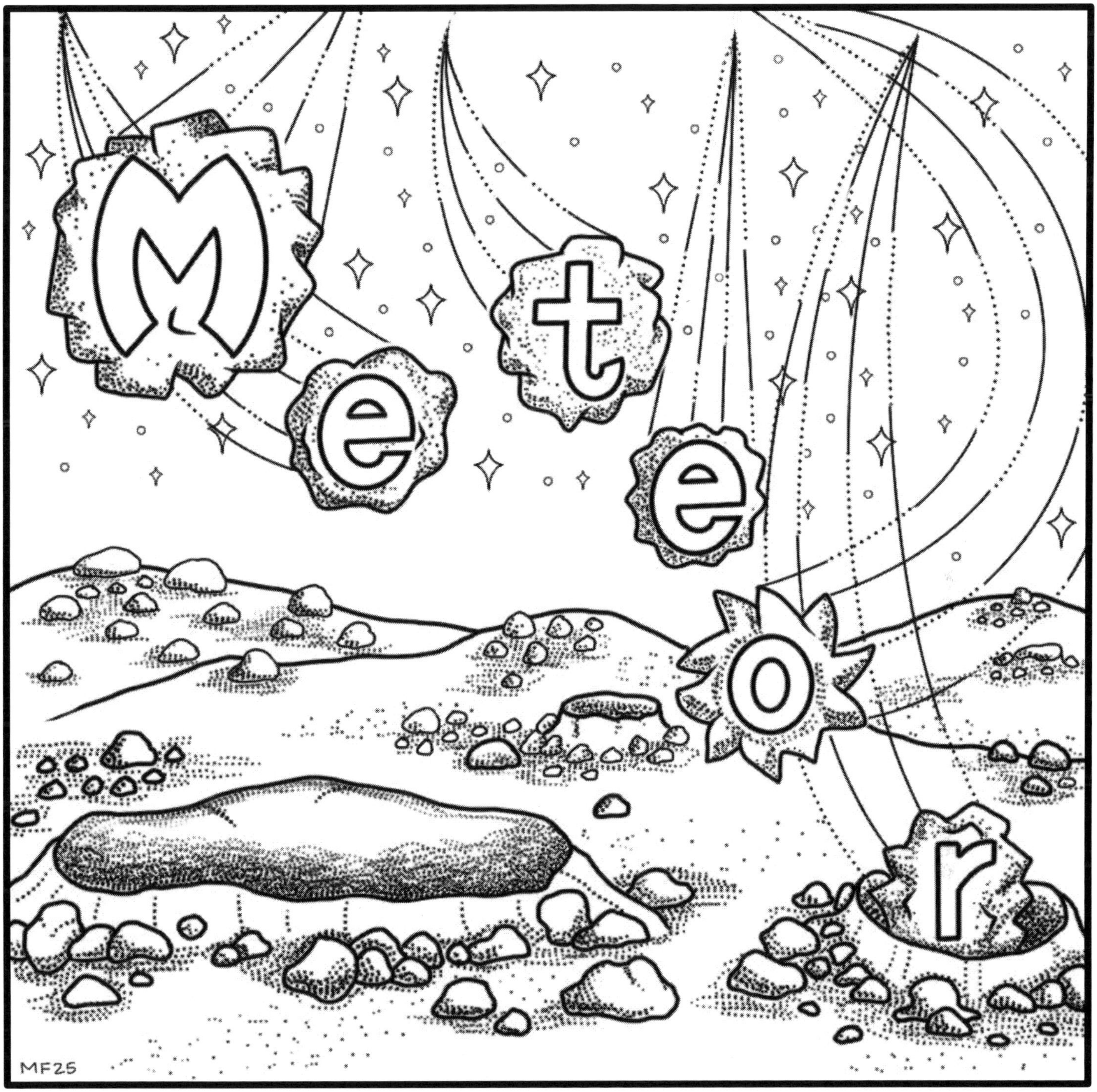

What is a meteor, how is it different from an asteroid?	What causes a meteor to glow or produce a "shooting star"?
What is a meteor shower, and when do they occur?	How can scientists learn about the early solar system by studying meteorites?

ALIEN LIFE

For thousands of years, humans have looked up at the night sky and wondered: Are we alone in the universe? With billions of galaxies, each containing billions of stars, the idea that Earth is the only place where life exists seems both extraordinary and uncertain. Scientists use the term extraterrestrial life to describe any form of life that may exist beyond Earth.

Life as we know it requires certain basic conditions: liquid water, a source of energy, and chemical building blocks such as carbon. On Earth, life thrives in oceans, deserts, deep underground, and even near volcanic vents on the seafloor. These discoveries have expanded our understanding of where life could survive, suggesting that alien life might not look like the creatures we imagine in movies. It could be microscopic, simple, and very different from plants or animals.

In recent decades, scientists have discovered that the basic building blocks of life, known as biomolecules, exist beyond Earth. Organic molecules such as amino acids and other carbon-based compounds have been found in meteorites that landed on Earth, in samples collected from asteroids, and in dust from comets. Laboratory analysis shows that these molecules can form naturally in space, even in the cold and radiation-filled environment between planets. Some scientists believe that impacts from meteorites and comets may have delivered these important ingredients to early Earth, helping to spark the development of life on our planet.

Scientists are actively searching for signs of life within our own solar system. Moons such as Europa (orbiting Jupiter) and Enceladus (orbiting Saturn) have oceans beneath their icy surfaces. Mars shows evidence of ancient rivers and lakes. These worlds may once have had, or may still have, environments suitable for life.

Beyond our solar system, astronomers have discovered thousands of exoplanets, planets orbiting distant stars. Some of these planets lie in the "habitable zone," where temperatures might allow liquid water to exist. Powerful telescopes, such as the James Webb Space Telescope, analyze the atmospheres of distant planets, searching for chemical signatures that could indicate biological activity.

Although no confirmed evidence of alien life has been found yet, the search continues. Whether life elsewhere is simple microbes or intelligent beings remains one of the greatest scientific questions of our time. Exploring this mystery helps us better understand our own planet, the origins of life, and our place in the vast universe.

What conditions are considered necessary for life as we know it, and why are these conditions important?	How do discoveries of biomolecules in meteorites, asteroids, and comet dust support the possibility of life elsewhere in the universe?
Why are moons such as Europa and Enceladus considered strong candidates in the search for extraterrestrial life?	What is the “habitable zone,” and how do scientists use it when searching for life on exoplanets?

COSMIC RAY

Cosmic rays are high-energy particles that travel through space at nearly the speed of light. Despite their name, cosmic rays are not rays of light; they are mostly subatomic particles, such as protons and atomic nuclei, that originate from energetic events in the universe. When cosmic rays reach Earth, they interact with the atmosphere, producing cascades of secondary particles that scientists can detect at the surface. Most cosmic rays—about 90 percent—are protons, which are the nuclei of hydrogen atoms. Around 9 percent are helium nuclei, and the remaining small fraction consists of heavier atomic nuclei such as carbon, iron, and other elements. A tiny portion are electrons and antimatter particles such as positrons. This composition closely resembles the elemental makeup of stars, linking cosmic rays to stellar processes.

The existence of cosmic rays was confirmed in 1912 by Austrian physicist Victor Hess. Hess carried radiation detectors aboard high-altitude balloon flights and discovered that radiation levels increased with altitude, proving the source came from space rather than Earth. This experiment marked the first confirmed discovery of cosmic rays and earned Hess the Nobel Prize in Physics in 1936. Earlier scientists had noticed unexplained radiation, but Hess's work provided the decisive evidence.

Cosmic rays are observed constantly bombarding Earth from all directions. They originate from multiple sources, including the Sun, exploding stars known as supernovae, rapidly spinning neutron stars, and distant active galaxies with supermassive black holes. Solar cosmic rays are generally lower in energy, while galactic and extragalactic cosmic rays can carry enormous energies far beyond anything produced in human-made particle accelerators.

Cosmic rays do not orbit the Sun or planets in stable paths like asteroids or comets. Instead, they travel through space in mostly straight lines, but their paths become curved and scattered by magnetic fields in space, including Earth's magnetic field. Because of this deflection, tracing a cosmic ray back to its exact source is difficult. Cosmic rays may travel for thousands or even millions of years across the galaxy before reaching Earth.

Scientists know about cosmic rays through their interactions with Earth's atmosphere. When a cosmic ray strikes atmospheric atoms, it produces a particle shower of secondary particles that spread toward the ground. These interactions reveal information about the cosmic ray's energy and composition, even though the original particle may never reach Earth's surface.

Cosmic rays are studied using ground-based detectors, balloon experiments, and space-based instruments. Large observatories use arrays of detectors spread over wide areas to measure particle showers. Satellites and space telescopes directly detect cosmic rays before they enter the atmosphere. Scientists also use cloud chambers, scintillation detectors, and magnetic spectrometers to analyze particle type, energy, and direction. Together, these methods allow researchers to study some of the most energetic phenomena in the universe.

Cosmic rays help scientists understand high-energy processes, stellar explosions, and the behavior of matter under extreme conditions. They also influence Earth by contributing to atmospheric ionization and radiation exposure at high altitudes. By studying cosmic rays, scientists gain insight into the most powerful events in the cosmos and the structure of our galaxy.

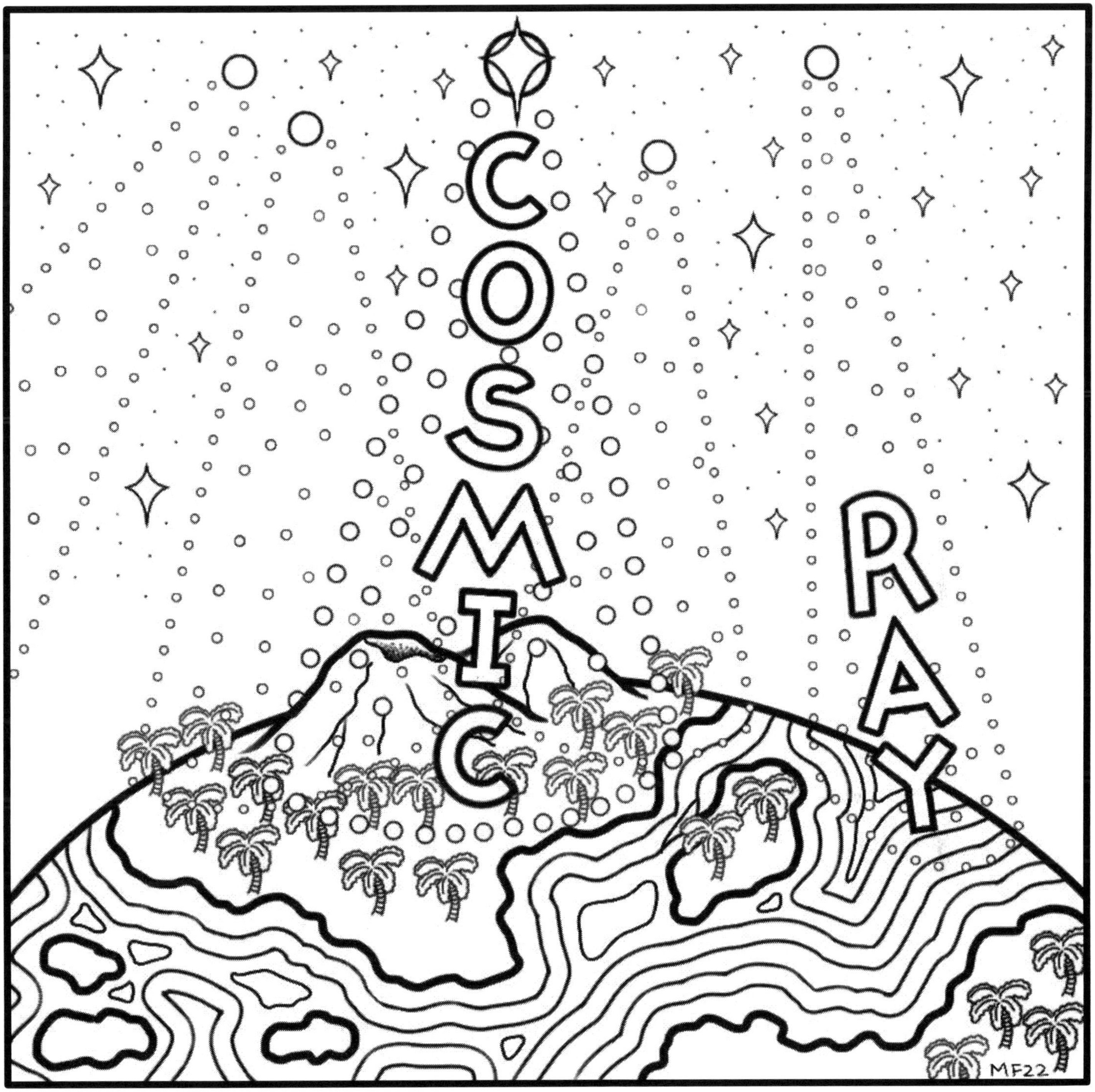

What are cosmic rays, and why are they not considered rays of light?

How did Victor Hess's balloon experiments help confirm the existence of cosmic rays?

Where do cosmic rays originate, and how do magnetic fields affect their paths through space?

Why are cosmic rays important to scientists studying high-energy events in the universe and their effects on Earth?

TELESCOPE

Have you ever had the opportunity to look through a telescope at the night sky, or at a scenic location? Telescopes can provide hours of entertainment but they are also important. A telescope is a scientific instrument used to observe distant objects in space by collecting and focusing electromagnetic radiation, such as visible light, infrared, or radio waves. In astronomy, telescopes allow scientists to study stars, planets, galaxies, and other celestial objects that are too far away to be seen clearly with the human eye. By gathering more light than the eye can, telescopes make faint and distant objects appear brighter and more detailed.

The history of telescopes began in the early 1600s with the invention of the optical telescope. One of the first astronomers to use a telescope for scientific study was Galileo Galilei. He used a refracting telescope, which bends light using lenses, to observe the Moon, discover moons orbiting Jupiter, and study the phases of Venus. These observations provided strong evidence that Earth was not the center of the universe and helped transform astronomy into a modern science.

As telescope technology advanced, reflecting telescopes were developed. Reflecting telescopes use mirrors instead of lenses to gather and focus light, allowing them to be much larger and more powerful. Larger telescopes can collect more light, making it possible to observe extremely distant and faint objects. Advances in materials, precision engineering, and computer control have greatly improved the clarity and accuracy of astronomical observations.

Modern telescopes are no longer limited to visible light. Astronomers now use telescopes that detect many types of electromagnetic radiation, including radio waves, infrared radiation, ultraviolet light, X-rays, and gamma rays. Space-based telescopes operate above Earth's atmosphere, which blocks or distorts some wavelengths of light. These telescopes allow scientists to study phenomena such as black holes, star formation, and the expansion of the universe with much greater detail.

Telescopes are important because they allow astronomers to explore the universe and understand how it formed and evolved. Through telescope observations, scientists have discovered exoplanets, measured the age of the universe, and studied galaxies billions of light-years away. As technology continues to improve, telescopes remain one of the most powerful tools for expanding knowledge of space and our place within it.

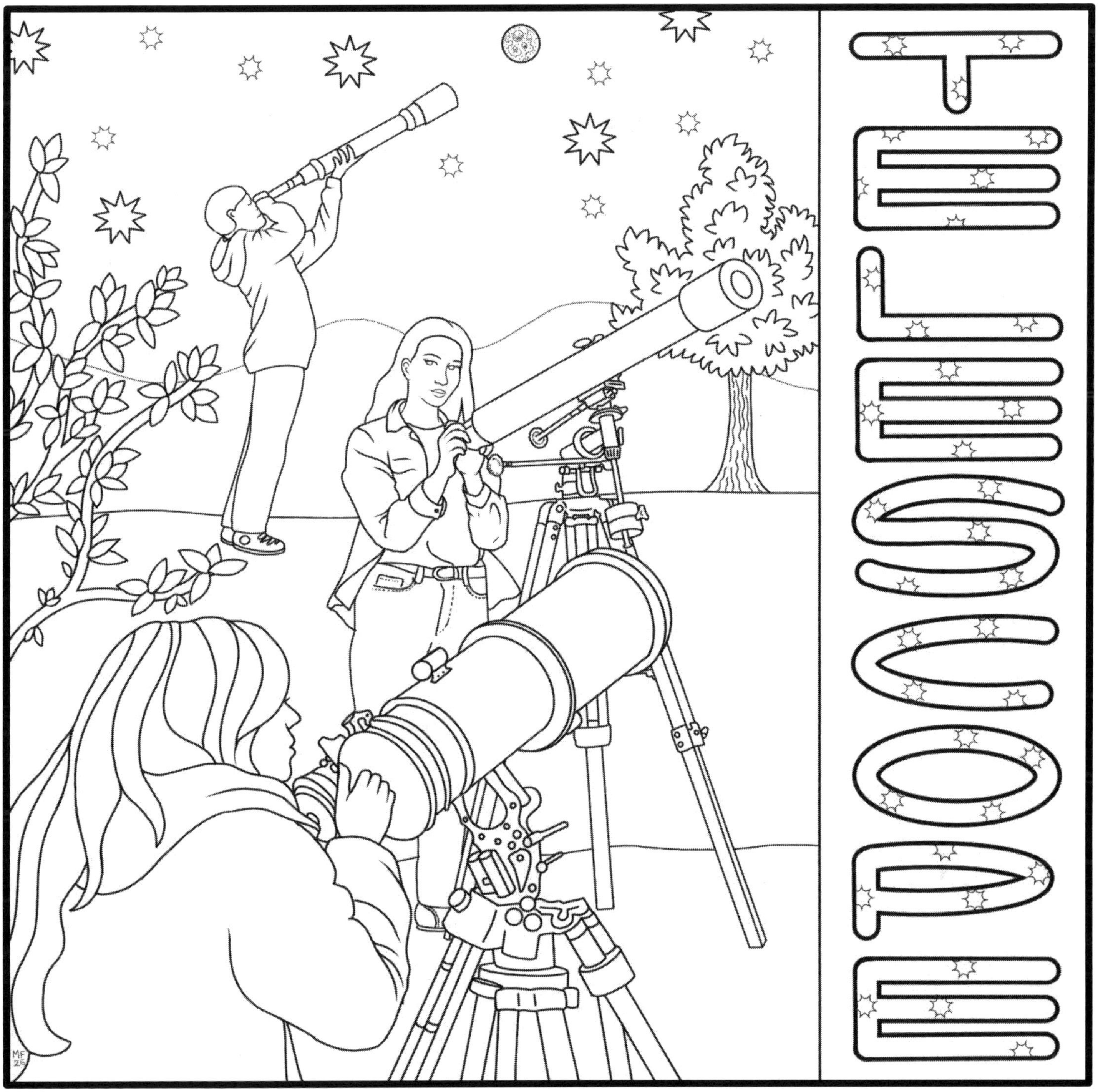

What is a telescope, and how does it help astronomers study space?

How did early telescopes used by astronomers like Galileo change our understanding of the universe?

What is one major difference between refracting and reflecting telescopes?

With modern telescopes designed to detect different types of electromagnetic radiation, which phenomenon have scientists been able to study?

SATELLITE

A satellite is any object that moves around a planet or another larger object due to gravity. Earth's gravity pulls satellites inward, while the satellite's forward motion keeps it moving sideways instead of falling straight down. This balance between gravity and forward motion causes the satellite to follow a curved path called an orbit. If a satellite moves too slowly, gravity will pull it back to Earth. If it moves too fast, it will escape Earth's gravity altogether. There are two main categories of satellites: natural satellites and artificial satellites. Natural satellites occur naturally and orbit planets due to gravity. Earth's Moon is an example of a natural satellite. Artificial satellites are human-made objects that are launched into space and placed into orbit around Earth or other celestial bodies. These satellites are carefully launched at precise speeds and angles, so they remain in orbit.

Artificial satellites serve many different purposes depending on their design and orbit. Communication satellites help transmit television signals, phone calls, and internet data across the globe. Weather satellites monitor clouds, storms, temperature, and atmospheric conditions, helping scientists predict weather and track hurricanes. Navigation satellites, such as those used for GPS, help determine location and guide transportation systems. Scientific satellites study space by observing stars, planets, and cosmic radiation, while Earth-observation satellites collect data about land, oceans, ice, and climate change.

Satellites play an important role in helping scientists, study both Earth and outer space. Earth-orbiting satellites allow scientists to observe changes in climate, track deforestation, measure sea levels, and monitor natural disasters such as wildfires and earthquakes. Space-based satellites can observe objects beyond Earth's atmosphere without interference from air or light pollution, making them essential tools for astronomy and space research.

The first artificial satellite launched into space was Sputnik 1, which was launched by the Soviet Union in 1957. Sputnik 1 was about the size of a beach ball and transmitted simple radio signals back to Earth. Its launch was significant because it proved that objects could be placed into orbit and marked the beginning of the Space Age. Sputnik's success led to rapid advancements in space technology and intensified the space race between nations.

Natural satellites, such as moons, have a powerful influence on the planets they orbit. Earth's Moon affects ocean tides through its gravitational pull and helps stabilize Earth's tilt, which contributes to a relatively stable climate. Some moons cause tidal heating, which can create volcanic activity, as seen on Jupiter's moon Io. Natural satellites can also influence a planet's rotation and play a role in shaping planetary systems over time.

Together, natural and artificial satellites help scientists understand gravity, motion, planetary systems, and Earth's changing environment. From guiding navigation systems to revealing distant galaxies, satellites are essential tools for exploring both our planet and the universe beyond it.

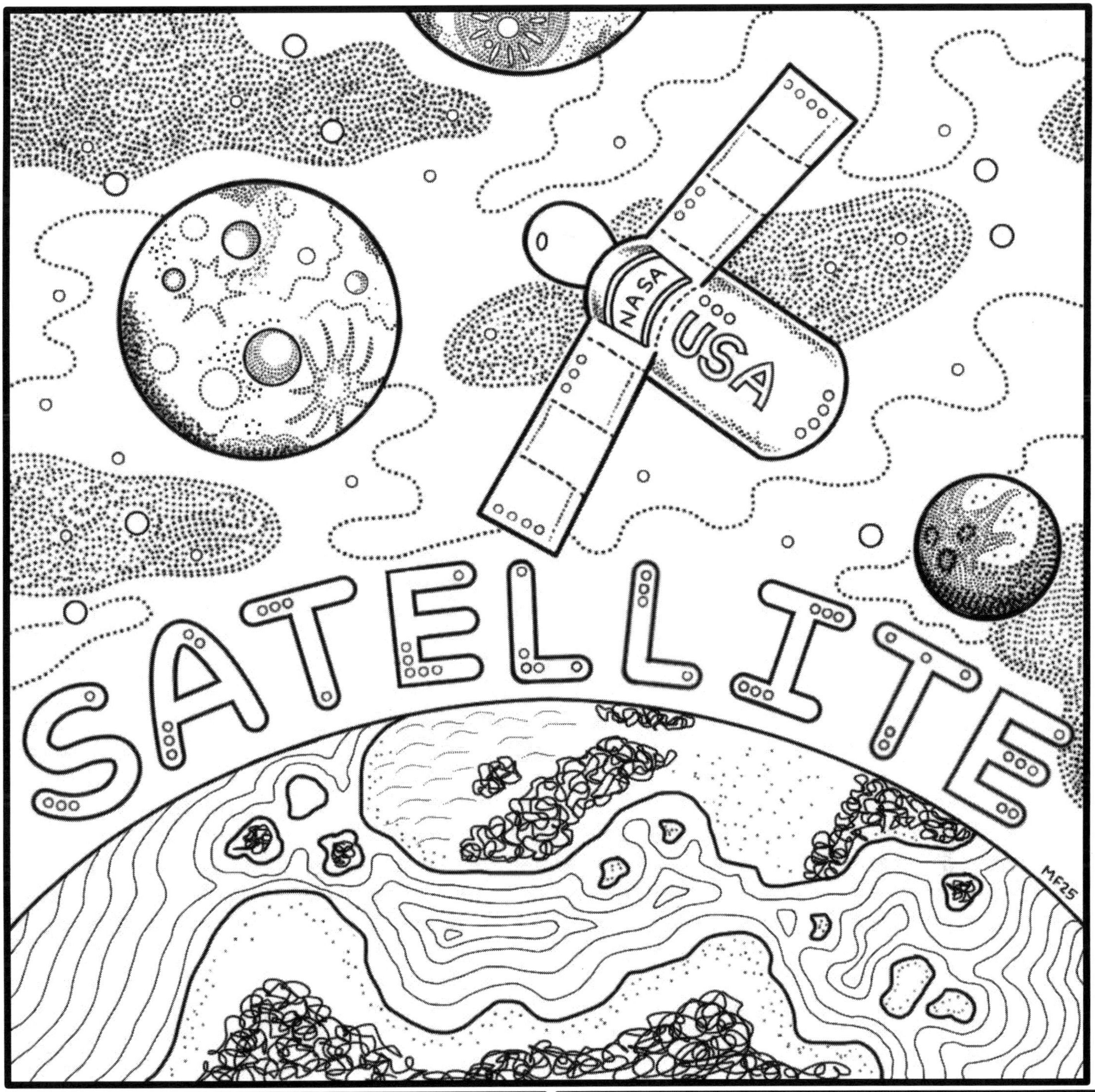

What is a satellite, and how does it stay in orbit around Earth?	What are the different types of satellites and their purposes?

How do satellites help us study space and Earth?	What was the first artificial satellite launched into space, and why was it significant?

ORBITER

An orbiter is a type of spacecraft designed to travel to a planet, moon, or other celestial body and move around it in a stable orbit. Unlike landers or rovers, orbiters do not land on the surface. Instead, they repeatedly circle their target, allowing scientists to observe it over long periods of time. Orbiters use gravity and precise speed to remain in orbit while collecting data from above.

Orbiters are important scientific tools because they provide a broad, long-term view of celestial objects. From orbit, these spacecraft can map surfaces, study atmospheric conditions, measure temperatures, analyze chemical composition, and observe changes over time. Orbiters are especially useful for studying weather patterns, seasonal changes, volcanic activity, and surface features that may not be visible from Earth-based telescopes.

The history of orbiters began during the early space age, shortly after the first artificial satellites were launched around Earth. As space technology improved, scientists began sending orbiters to other planets and asteroids. Early planetary orbiters allowed astronomers to gather the first detailed images and measurements of planets such as Mars and Venus. These missions revealed surface features, cloud layers, and atmospheric conditions that were previously unknown.

Advances in technology have greatly improved the capabilities of modern orbiters. Today's orbiters are equipped with high-resolution cameras, spectrometers, radar systems, and communication tools that can transmit large amounts of data back to Earth. Improved computer systems and navigation technology allow orbiters to adjust their paths, enter precise orbits, and operate for many years. Some orbiters work together with landers and rovers by acting as communication relays between the surface and Earth.

Orbiters play a vital role in space exploration. They help scientists choose safe landing sites, monitor planetary environments, and compare different locations across the solar system. By remaining in orbit for long periods, orbiters provide continuous data that deepens understanding of planetary systems and supports future exploration missions.

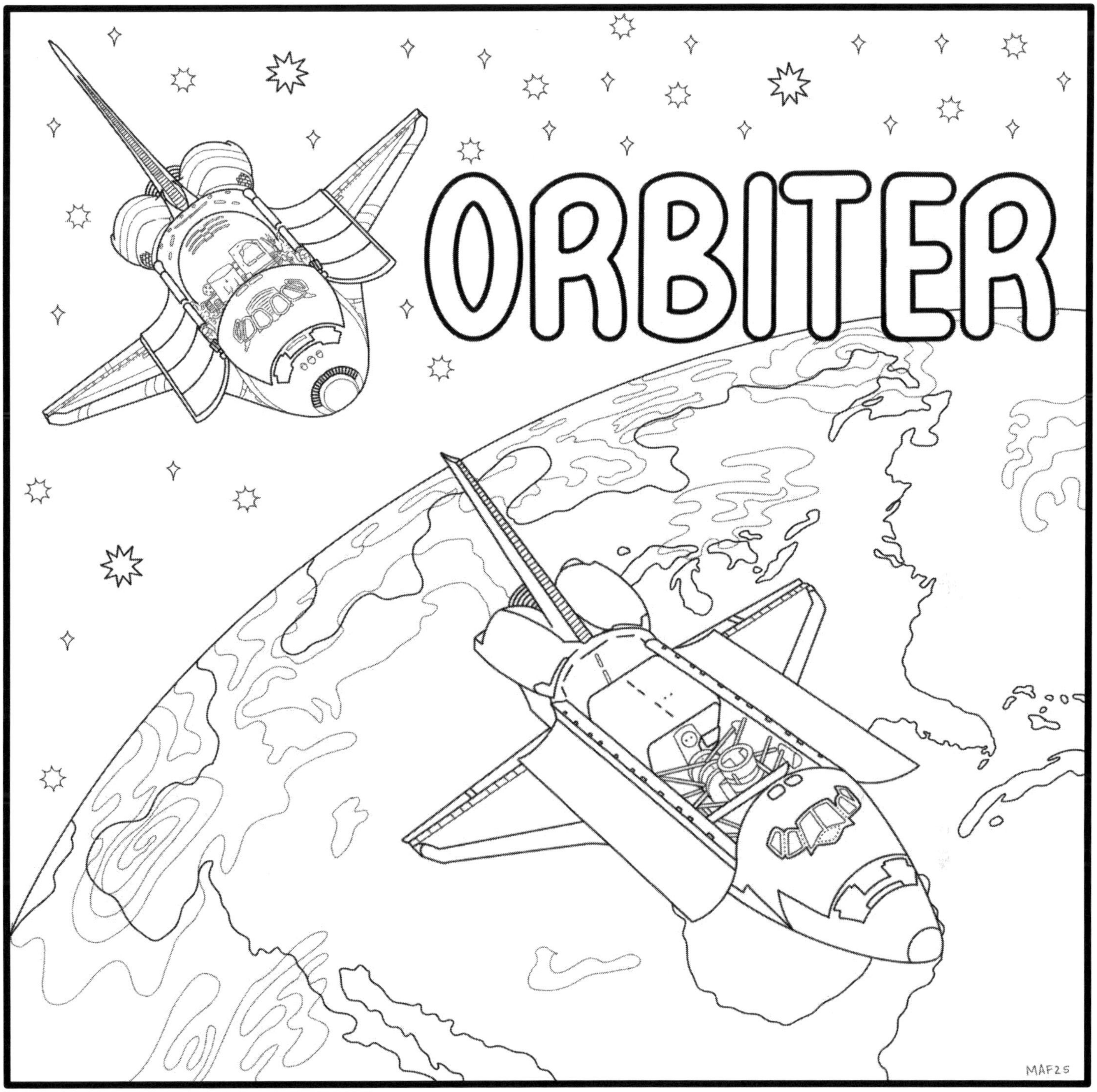

What is an orbiter, and how does it differ from a lander or rover?	Why are orbiters useful for studying planets and moons over long periods of time?

How did early orbiters change what scientists knew about other planets?	What technological advances have improved the effectiveness of modern orbiters?

ROVER

Have you ever wondered what it would be like to walk on the surface of Mars or Venus? Have you ever wondered how scientists know about the surface of different planets? Rovers are robotic vehicles designed to explore the surface of planets, moons, or other celestial bodies. Unlike spacecraft that orbit above a planet or landers that remain stationary, rovers can move across terrain, allowing them to study multiple locations during a single mission. Rovers are controlled remotely by scientists and engineers on Earth and are equipped with cameras, sensors, and scientific instruments that collect data and send it back to researchers.

One of the most important roles of rovers is to study environments that are too dangerous or distant for humans. Rovers can analyze rocks and soil, measure temperature and atmospheric conditions, and search for signs of water or past life. By traveling across a planet's surface, rovers provide detailed information about geology, climate, and surface features that cannot be gathered from orbit alone. This information helps scientists better understand how planets form and change over time.

The history of rovers began during the early years of space exploration. In 1971, the Soviet Union successfully operated Lunokhod 1, the first rover to explore the surface of the Moon. Later, NASA sent rovers to Mars, beginning with Sojourner in 1997. Sojourner demonstrated that a small robotic vehicle could move, navigate, and perform experiments on another planet. This success led to more advanced rovers such as Spirit, Opportunity, Curiosity, and Perseverance, each designed with improved mobility, stronger communication systems, and more sophisticated scientific tools.

Modern rovers are highly advanced machines capable of operating for years in harsh environments. They use solar panels or nuclear power sources for energy and can make limited decisions on their own to avoid obstacles. Some rovers are also designed to collect samples that may one day be returned to Earth. These missions are especially important for preparing future human exploration by identifying safe landing sites and available resources such as water ice.

Rovers are important because they extend human exploration beyond Earth without putting astronauts at risk. They act as robotic scientists, traveling across alien landscapes and sending back images and data that deepen our understanding of the solar system. Through rover missions, scientists have discovered evidence that Mars once had liquid water, reshaping ideas about the possibility of life beyond Earth.

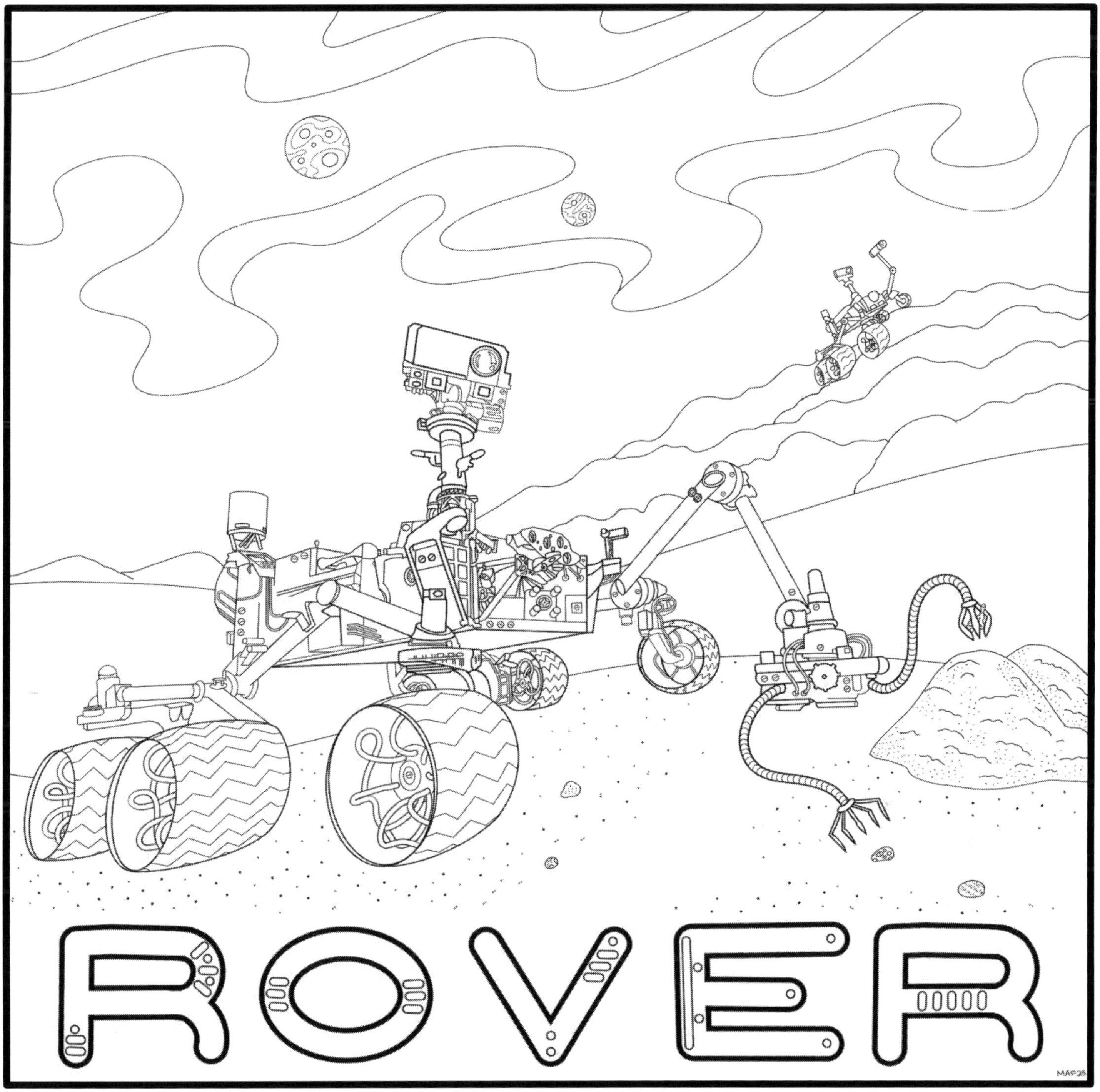

What makes rovers different from orbiters and stationary landers?	Why are rovers especially useful for exploring planets like Mars?
Describe one historical rover mission and explain why it was important.	How do rover missions help prepare for future human space exploration?

SPACE LANDER

Space landers are spacecraft designed to travel through space and make a controlled landing on the surface of a planet, moon, or other celestial body. Unlike orbiters, which remain in space, or rovers, which move across the surface, landers stay in one location after touchdown. Their main purpose is to collect close-up data by directly interacting with the surface, atmosphere, or environment of the object they land on.

One of the most important roles of space landers is scientific observation and measurement. Landers carry instruments that can analyze soil and rock samples, measure temperature and atmospheric pressure, detect radiation, and study chemical composition. Because landers operate directly on the surface, they can gather detailed information that cannot be obtained from orbit alone. This data helps scientists understand planetary conditions, geological history, and whether an environment could support life.

The history of space landers began during the early years of space exploration. In the 1960s and 1970s, the United States and the Soviet Union developed landers to explore the Moon and other planets. NASA's Apollo missions used lunar landers to safely deliver astronauts to the Moon's surface and return them to orbit. Robotic landers such as the Viking landers later touched down on Mars in the 1970s, becoming the first successful missions to conduct experiments directly on the Martian surface.

As technology advanced, landers became more sophisticated and capable of reaching more distant and challenging destinations. Modern landers have successfully explored places such as Mars, asteroids, and comets. Some landers are designed to study surface chemistry, while others focus on seismic activity or atmospheric conditions. Advanced landing systems, including retrorockets, airbags, and precision navigation, allow landers to touch down safely in difficult terrain.

Space landers are important because they allow scientists to study other worlds up close without sending humans into dangerous environments. They provide essential information about planetary surfaces and conditions, support future missions by identifying safe landing areas, and help answer key questions about the formation of the solar system and the possibility of life beyond Earth. By remaining stationary and performing long-term measurements, landers play a crucial role in expanding our understanding of space.

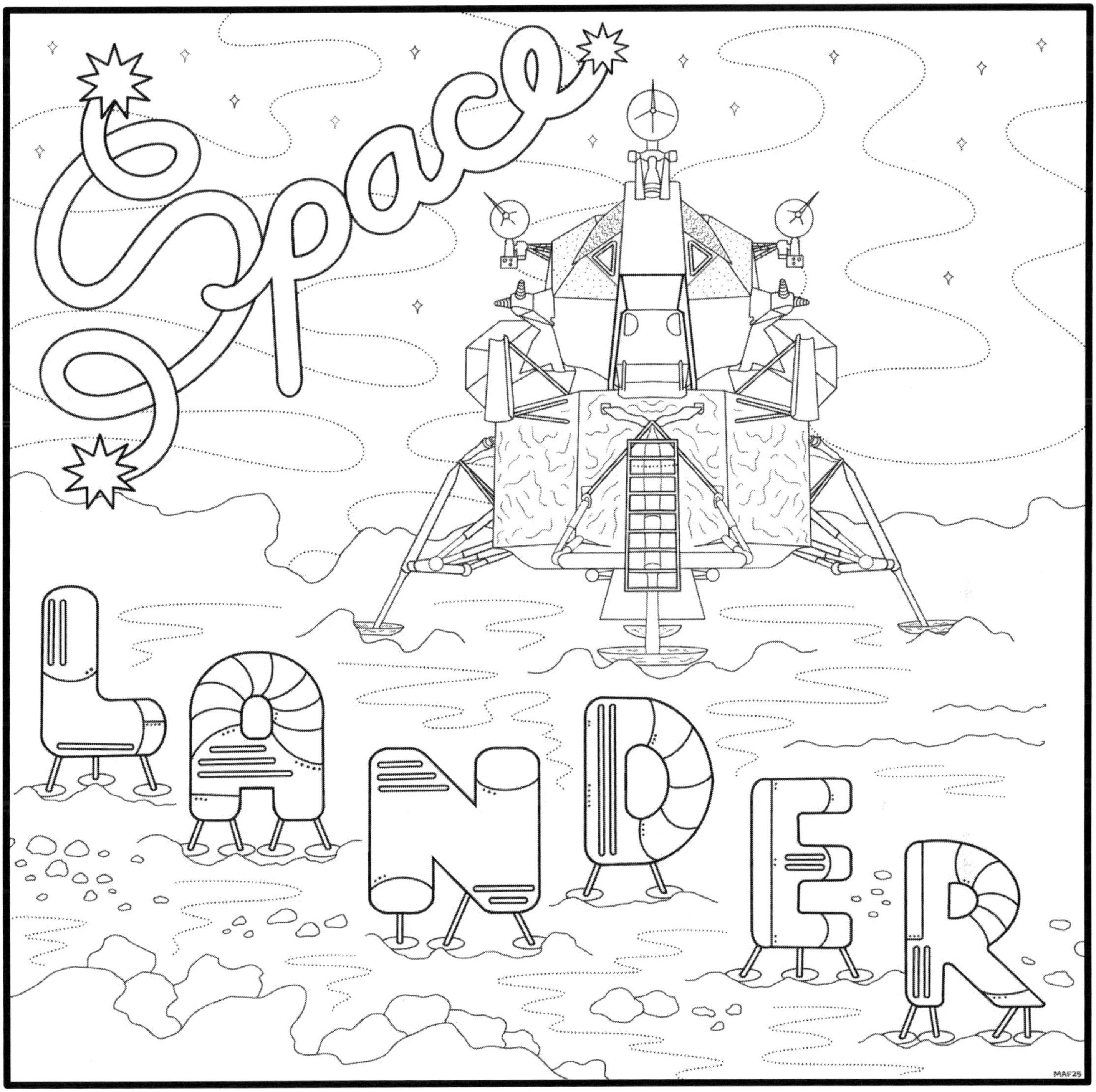

What is the primary purpose of a space lander?	How do space landers collect information that orbiters cannot?
Describe one historical space lander mission and explain why it was important.	Why are space landers valuable for future space exploration?

SPACE PROBE

A space probe is an unmanned spacecraft designed to travel far beyond Earth to study planets, moons, asteroids, comets, and deep space. Unlike crewed spacecraft, space probes do not carry astronauts. Instead, they are equipped with scientific instruments, cameras, and sensors that collect data and transmit it back to Earth. Space probes can fly past objects, orbit them, or even land on their surfaces, depending on the mission's goals.

Space probes are important because they allow scientists to explore regions of space that are too distant, dangerous, or costly for human missions. Probes can measure temperature, radiation, magnetic fields, gravity, and chemical composition. They also capture detailed images and send back information that helps scientists understand how planets form, how solar systems evolve, and what conditions exist beyond Earth. Many discoveries about the solar system have come directly from data collected by space probes.

The history of space probes began during the early space race. In the late 1950s and early 1960s, the United States and the Soviet Union launched the first probes toward the Moon and nearby planets. One of the earliest successful space probes was Sputnik 1, which demonstrated that objects could orbit Earth. Later probes such as Mariner missions explored Venus and Mars, providing the first close-up images of other planets and revealing that they were very different from Earth.

As technology improved, space probes were sent farther into the solar system and beyond. Probes like Voyager 1 and Voyager 2 explored the outer planets and continue traveling through interstellar space today. Other probes have landed on asteroids, studied comets, and orbited distant planets for years at a time. Modern space probes are designed to operate for long periods, withstand extreme conditions, and make precise measurements that help answer questions about the origins of the solar system and the possibility of life elsewhere in the universe.

Space probes are essential tools for space exploration because they expand human knowledge without risking human life. By acting as robotic explorers, they gather information that guides future missions, improves scientific models, and deepens our understanding of the universe. From nearby planets to the edge of the solar system, space probes continue to push the boundaries of discovery.

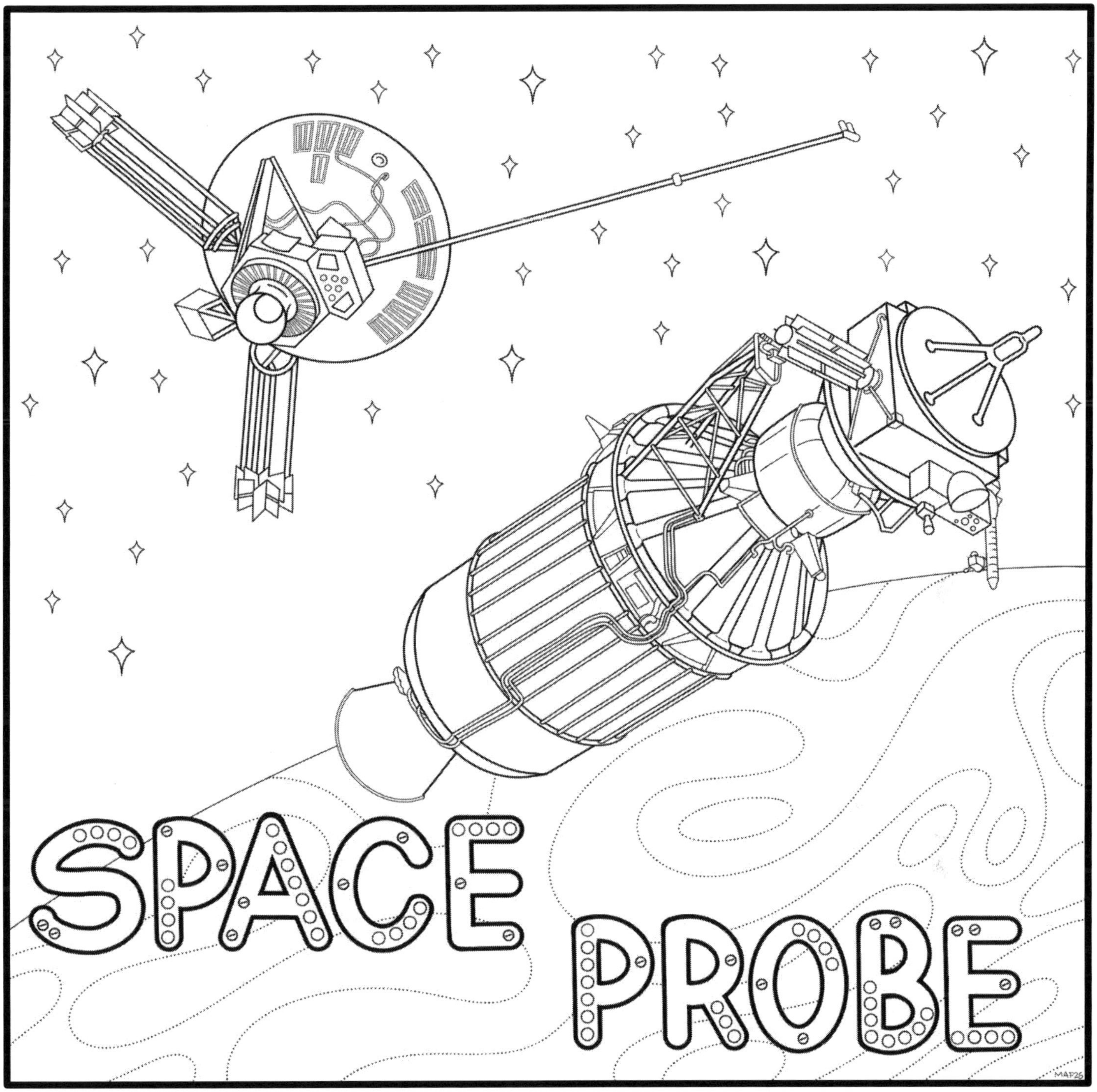

What is a space probe, and how is it different from a crewed spacecraft?	Why are space probes important for exploring distant parts of the solar system?
Describe one early space probe mission and explain why it was significant.	How have modern space probes improved our understanding of the universe?

HISTORY OF ASTRONAUTS

Have you ever dreamed of floating in space, visiting different galaxies, and floating among the stars? Before humans could safely travel into space, scientists needed to understand how living organisms would respond to microgravity, intense acceleration, and cosmic radiation. For this reason, the first space travelers were not humans.

In 1947, the United States launched fruit flies aboard a V-2 rocket to study the effects of radiation exposure. These insects were chosen because their genetic structure is similar to humans and their short life cycle allowed scientists to observe changes quickly. After the success of these early experiments, animals such as mice, monkeys, and dogs were sent into space to gather more detailed biological data. One of the most well-known early space missions involved Laika, a dog launched by the Soviet Union aboard Sputnik 2 in 1957. Laika's mission provided valuable information about heart rate, breathing, and stress during launch and orbit, even though the technology at the time did not allow for a safe return. These animal missions demonstrated that living organisms could survive the conditions of spaceflight, which was a critical step toward sending humans beyond Earth.

As scientists learned more about how life responded to space, engineers worked to develop the technology needed to support human astronauts. This included creating pressurized spacecraft, reliable life-support systems that supplied oxygen and removed carbon dioxide, and heat shields capable of protecting spacecraft during re-entry into Earth's atmosphere. Engineers also developed more powerful and stable rockets that could overcome Earth's gravity and place spacecraft into orbit. These advancements made it possible for humans to survive launch, remain alive in space, and return safely to Earth.

During the Cold War, two major space programs emerged and advanced human spaceflight along parallel paths. In the United States, NASA led astronaut missions, while the Soviet Union developed its own program, whose space travelers were known as cosmonauts. Although the terminology differed, astronauts and cosmonauts performed similar roles and faced the same physical challenges of space travel.

In April 1961, Soviet cosmonaut Yuri Gagarin became the first human to travel into space and orbit Earth aboard Vostok 1. His successful mission proved that humans could survive extended periods in orbit. Shortly afterward, American astronaut Alan Shepard completed the first U.S. human spaceflight on a suborbital mission. Both early astronauts and cosmonauts were typically military test pilots or engineers with extensive flight experience, strong technical knowledge, and exceptional physical fitness. The Soviet space program continued to reach major milestones, including the launch of Valentina Tereshkova in 1963. She became the first woman in space and the only woman to complete a solo space mission. Although the United States and Soviet Union competed intensely during this period, their combined achievements greatly expanded scientific understanding of human endurance, orbital motion, and long-duration spaceflight.

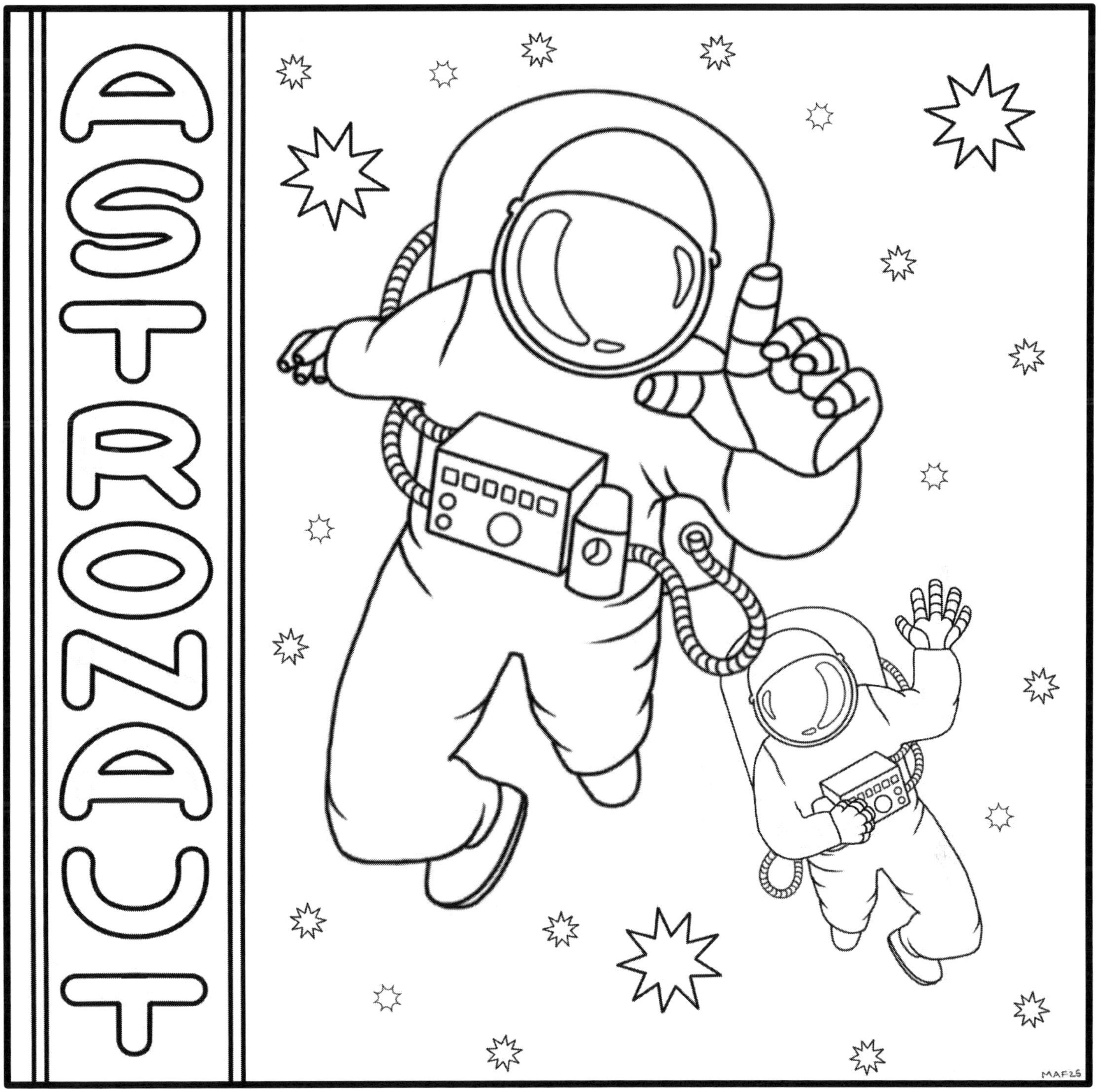

What is an astronaut?

Which organisms were sent into space before humans?

What types of technology needed to be developed to allow people to go into space?

Which two countries led Earth in space exploration?

ASTRONAUT TRAINING

As space missions grew longer and more complex, astronaut training became increasingly specialized. Modern astronauts and cosmonauts train for years before being assigned to a mission. Their training includes physical conditioning to prepare their bodies for the intense forces experienced during launch and re-entry. They also train in underwater facilities to simulate the weightless conditions of space, allowing them to practice spacewalks and equipment repairs. Flight training helps astronauts develop quick decision-making skills and spatial awareness, while survival training prepares them for emergency landings in environments such as oceans, deserts, or cold regions.

A spacewalk is when an astronaut leaves a spacecraft and works outside in space. The official term for a spacewalk is extravehicular activity (EVA), meaning "activity outside the vehicle." During a spacewalk, astronauts wear specially designed spacesuits that provide oxygen, regulate temperature, protect them from radiation and micrometeoroids, and allow them to communicate with mission control. Spacewalks are performed to repair spacecraft, install new equipment, conduct scientific experiments, or test new technology. Because space has no air, extreme temperatures, and microgravity, astronauts must be carefully trained and always remain tethered to the spacecraft for safety.

In addition to physical preparation, astronauts receive extensive technical training. They learn how to operate spacecraft systems, conduct scientific experiments, and use robotic arms and tools aboard space stations. Because astronauts live and work together in confined spaces for long periods, teamwork, communication, and problem-solving skills are essential parts of their preparation.

Astronauts return to Earth inside a spacecraft that is specially designed to survive reentry through the atmosphere. First, the spacecraft fires its engines in a deorbit burn, slowing it down so Earth's gravity pulls it out of orbit. As it reenters the atmosphere, friction with air molecules creates extreme heat, forming a glowing plasma around the capsule. A heat shield protects the astronauts from temperatures that can exceed 1,600°C (3,000°F). As the spacecraft slows further, parachutes deploy to reduce its speed before landing in the ocean or on land, depending on the vehicle. Recovery teams then retrieve the capsule and assist the astronauts, who may feel temporarily weak because their bodies have adjusted to microgravity during their time in space.

Becoming an astronaut is a highly competitive process. Most space agencies require candidates to hold at least a bachelor's degree in science, engineering, mathematics, or medicine. Many astronauts also have advanced degrees or years of professional experience in fields such as research, aviation, or healthcare. Candidates must meet strict physical and medical standards and demonstrate the ability to remain calm and effective under pressure. Today's astronauts are no longer only pilots. Many are scientists, engineers, doctors, and educators who conduct experiments aboard the International Space Station and help prepare for future missions to the Moon, Mars, and beyond. From fruit flies and dogs to international crews living in orbit, the evolution of astronauts reflects the ongoing partnership between humans and technology. As space exploration continues to advance, astronauts and cosmonauts will remain at the forefront of discovery, adaptation, and innovation beyond Earth

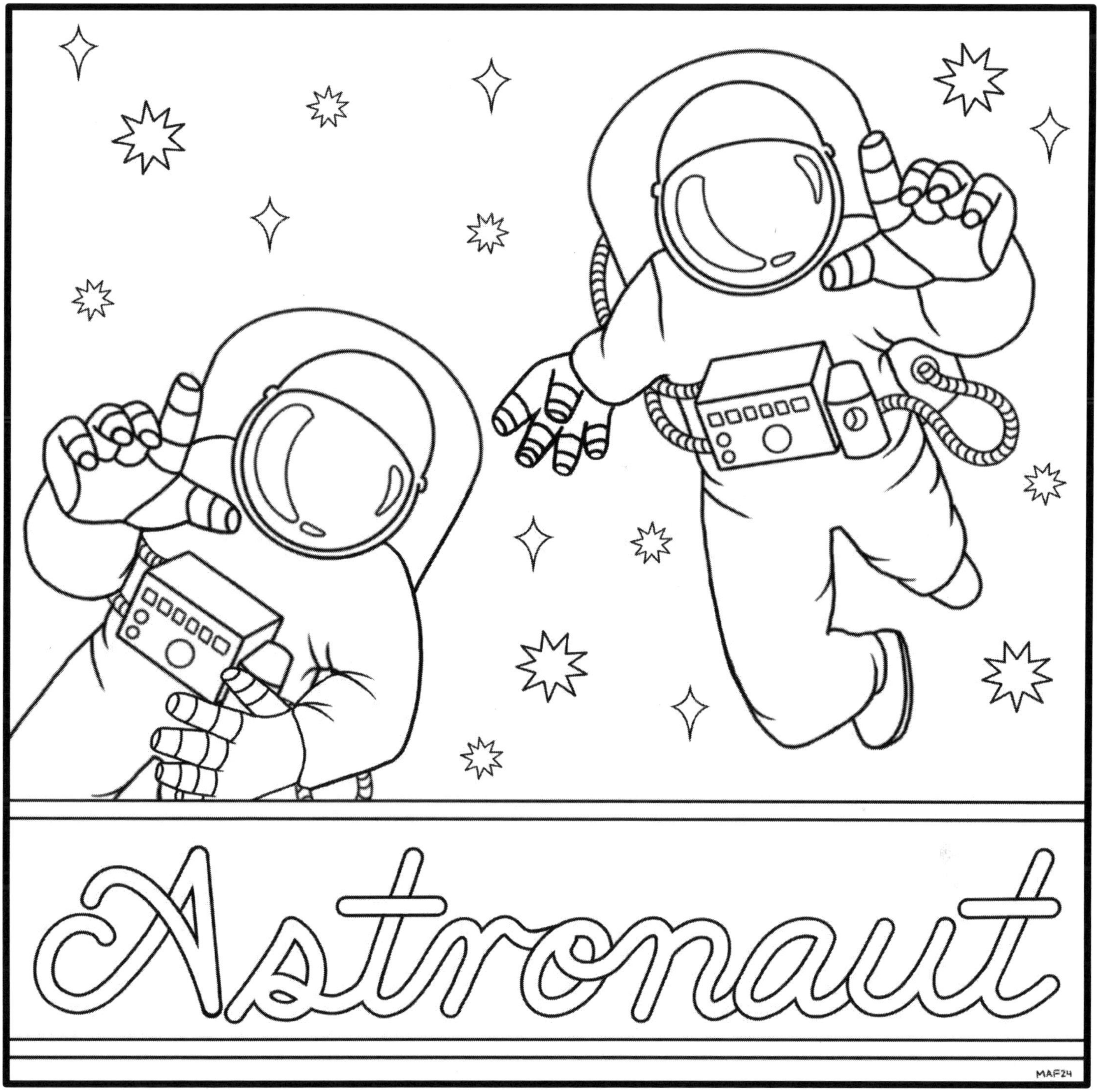

Which professional backgrounds do astronauts have?	What types of training do astronauts complete before going into space?
What is a spacewalk, and why is it important?	How do astronauts return safely to Earth after a mission?

CAPSULE

Spacecraft capsules are small, enclosed vehicles designed to carry astronauts, equipment, or scientific samples into space and safely return them to Earth. Unlike space shuttles or spaceplanes, capsules are compact, cone-shaped or spherical vehicles that rely on heat shields, parachutes, and precise engineering to survive launch, orbit, and re-entry. Their primary purpose is to protect human life and sensitive instruments during the most dangerous phases of space travel.

One of the most important roles of a spacecraft capsule is crew safety. During launch, capsules must withstand extreme acceleration and vibration. While in space, they provide a sealed, pressurized environment that supplies oxygen, removes carbon dioxide, and protects astronauts from temperature extremes and radiation. When returning to Earth, capsules face intense friction as they pass through the atmosphere. Heat shields absorb and dissipate this energy, preventing the capsule from burning up before parachutes slow it for a safe landing.

The history of spacecraft capsules dates back to the early days of human spaceflight. The United States used capsules in the Mercury and Apollo programs, while the Soviet Union developed capsule-based spacecraft such as Vostok. The first human spaceflight, completed by Yuri Gagarin in 1961, used the Vostok capsule. Later, Apollo capsules carried astronauts to the Moon and back, demonstrating that capsules could support long-duration missions and safely return crews from deep space.

After the Apollo era, capsule design continued to evolve. Modern capsules such as those used by NASA and commercial space companies are designed with improved materials, digital control systems, and reusable components. These capsules are now used to transport astronauts to and from the International Space Station and are being developed for future missions to the Moon and Mars. Despite advances in space technology, capsules remain a preferred design because of their reliability, simplicity, and proven safety record.

Spacecraft capsules are important because they represent one of the safest and most efficient ways to send humans into space and bring them home. Their continued use shows how early engineering solutions laid the foundation for modern space exploration. From the first human orbit to current international missions, spacecraft capsules have played a critical role in expanding humanity's reach beyond Earth.

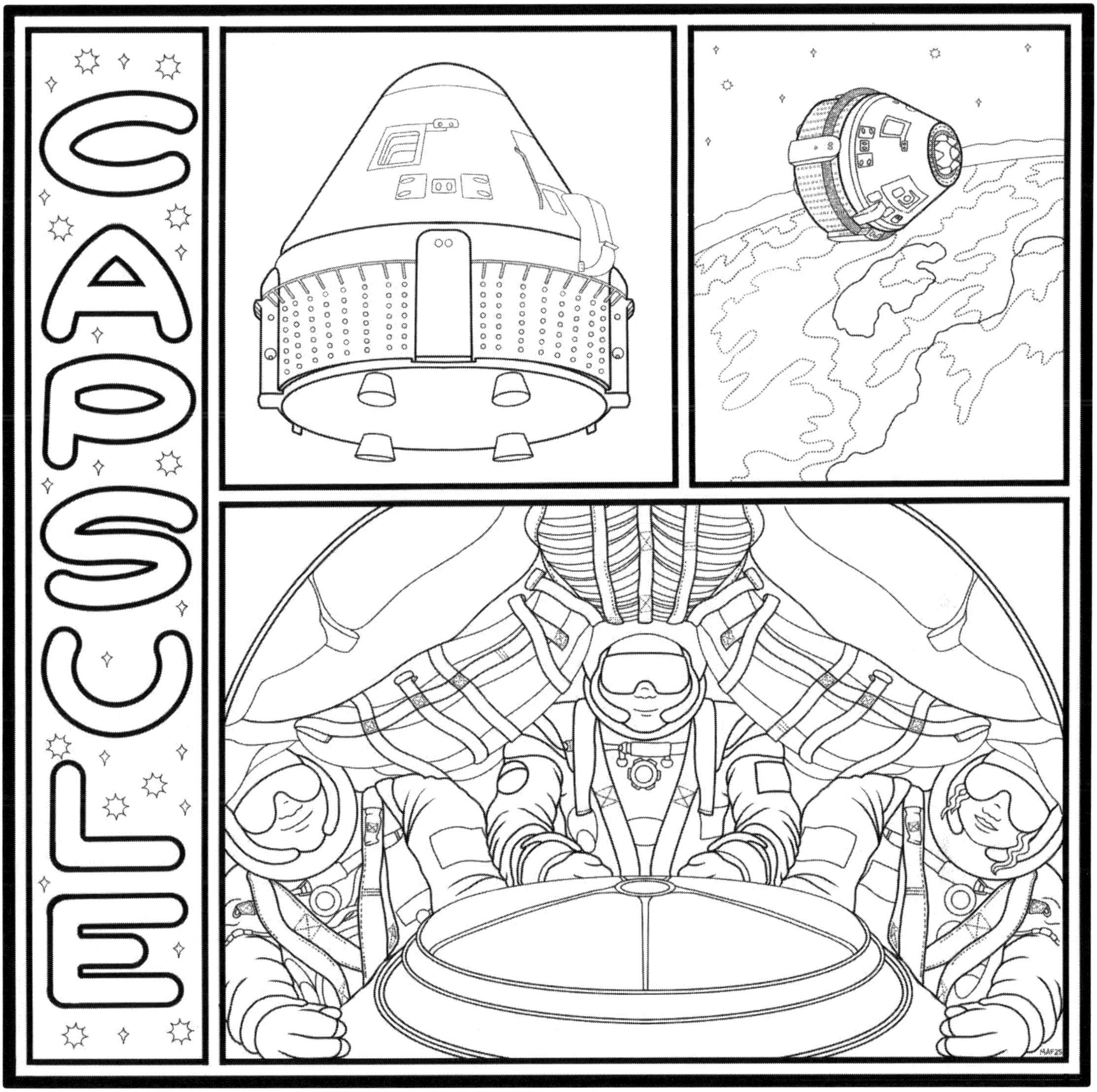

What is the primary purpose of a spacecraft capsule during space missions?	How do spacecraft capsules protect astronauts during re-entry into Earth’s atmosphere?
Name one early space program that used spacecraft capsules and explain why it was significant.	How have capsules improved over time?

SPACE STATION

A space station is a large spacecraft designed to remain in orbit around Earth for long periods of time while astronauts live and work aboard it. Unlike capsules or shuttles, space stations do not travel to distant destinations and return quickly. Instead, they serve as long-term laboratories where scientists can conduct experiments in the unique environment of microgravity. Space stations are carefully designed to support human life by providing oxygen, water, temperature control, and protection from the harsh conditions of space.

One of the most important purposes of a space station is scientific research. In microgravity, everyday forces such as gravity-driven settling and convection are greatly reduced. This allows scientists to study how materials, fluids, plants, and even the human body behave without gravity's influence. Research conducted on space stations has improved understanding of muscle loss, bone density, fluid movement, and the effects of long-term space travel on human health. These studies help prepare astronauts for future missions to the Moon and Mars while contributing to medical and technological advances on Earth.

The history of space stations began during the Cold War, when both the United States and the Soviet Union sought to maintain a human presence in space. In 1971, the Soviet Union launched Salyut 1, the world's first space station. It was followed by additional Salyut stations and later by Mir, which allowed astronauts to live in space for months at a time. The United States developed Skylab in the 1970s, which provided valuable information about long-duration spaceflight and solar observation.

Today, the most well-known space station is the International Space Station, or ISS, which is a cooperative project involving multiple countries. The ISS has been continuously occupied by astronauts since the year 2000. Crews aboard the station conduct experiments, test new technologies, and observe Earth from space. The ISS also serves as a model for international cooperation and has helped scientists learn how humans can live and work safely in space for extended periods.

Space stations are important because they allow continuous scientific research in space while supporting human exploration. They act as testing grounds for future missions, help improve life-support systems, and expand knowledge of how living organisms respond to long-term space conditions. By maintaining a permanent human presence in orbit, space stations play a critical role in advancing space science and exploration.

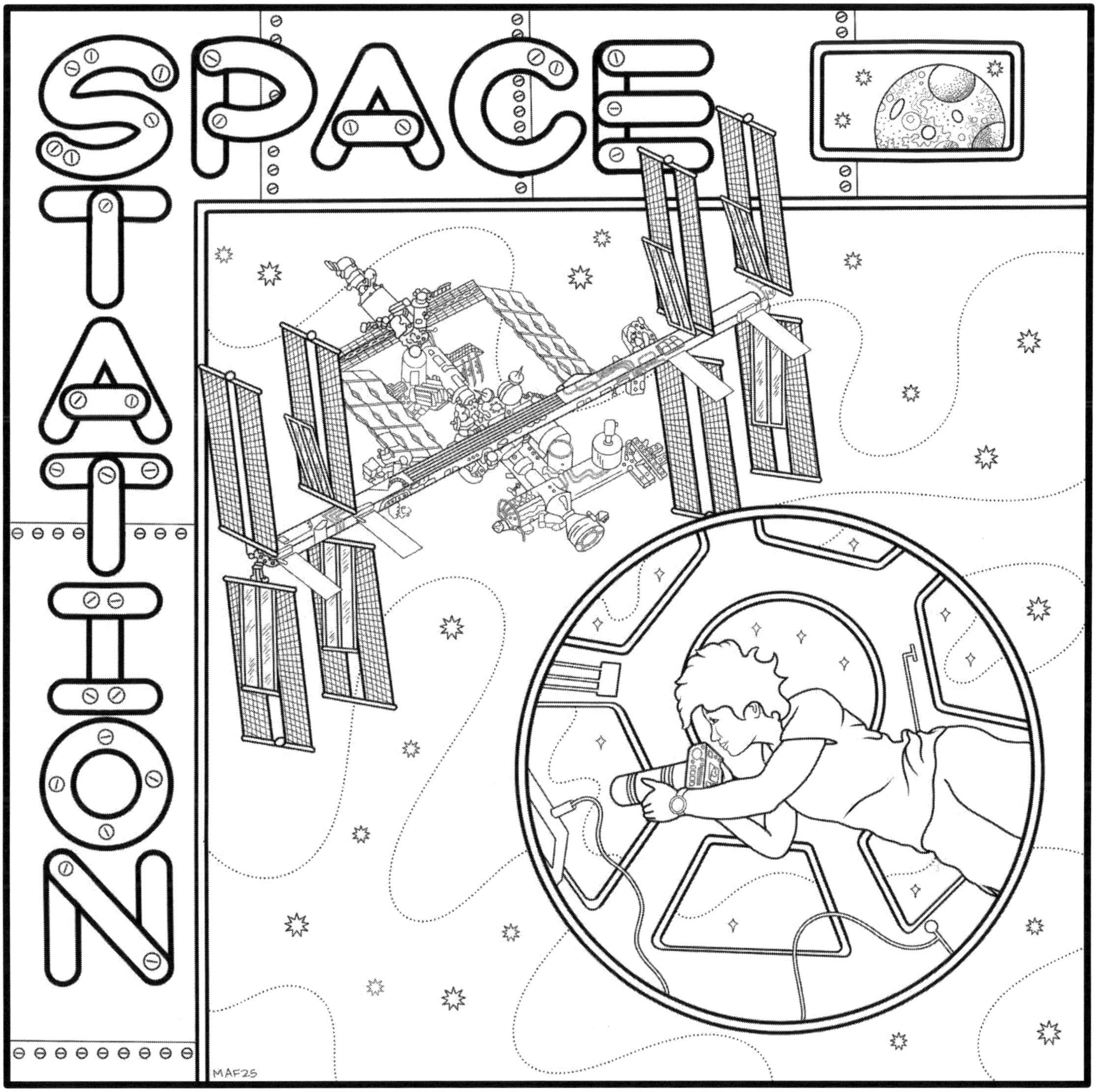

What is a space station, and how is it different from other types of spacecraft?	Why is microgravity an important environment for scientific research?
Describe one early space station and explain why it was significant.	How do space stations help prepare humans for future space missions?

SPACE SHUTTLE

A space shuttle is a reusable spacecraft designed to travel from Earth into orbit and return safely to the surface. Unlike capsules, which land using parachutes, space shuttles were built to glide back to Earth and land on a runway like an airplane. The main purpose of the space shuttle was to transport astronauts, satellites, and scientific equipment to and from space while allowing repeated use of the same vehicle, which saves money and reduces waste.

One of the most important features of the space shuttle was its ability to carry large payloads. The shuttle included a cargo bay that could hold satellites, space station components, and scientific experiments. Astronauts aboard the shuttle could deploy satellites into orbit, repair damaged equipment, and conduct experiments in microgravity. The shuttle also played a key role in assembling and servicing space stations, making it a versatile tool for space exploration.

The history of the space shuttle began in the 1970s when the United States sought a reusable spacecraft to reduce the cost of space travel. The first space shuttle mission launched in 1981, marking a new era of human spaceflight. Over the next three decades, the shuttle program completed more than 130 missions. These missions included launching major satellites, conducting scientific research, and helping build the International Space Station. The program was operated by NASA and represented a major shift from single-use spacecraft to reusable systems.

Despite its successes, the space shuttle program also faced serious challenges. Two shuttle disasters resulted in the loss of crews, highlighting the risks of human spaceflight. These tragedies led to improvements in safety and a deeper understanding of spacecraft design and mission planning. In 2011, the space shuttle program was officially retired, and the remaining shuttles were placed in museums for public display.

Space shuttles are important because they demonstrated the possibilities and challenges of reusable spacecraft. They advanced technology, expanded scientific knowledge, and helped establish a long-term human presence in space. The lessons learned from the space shuttle program continue to influence modern spacecraft design and future missions beyond Earth orbit.

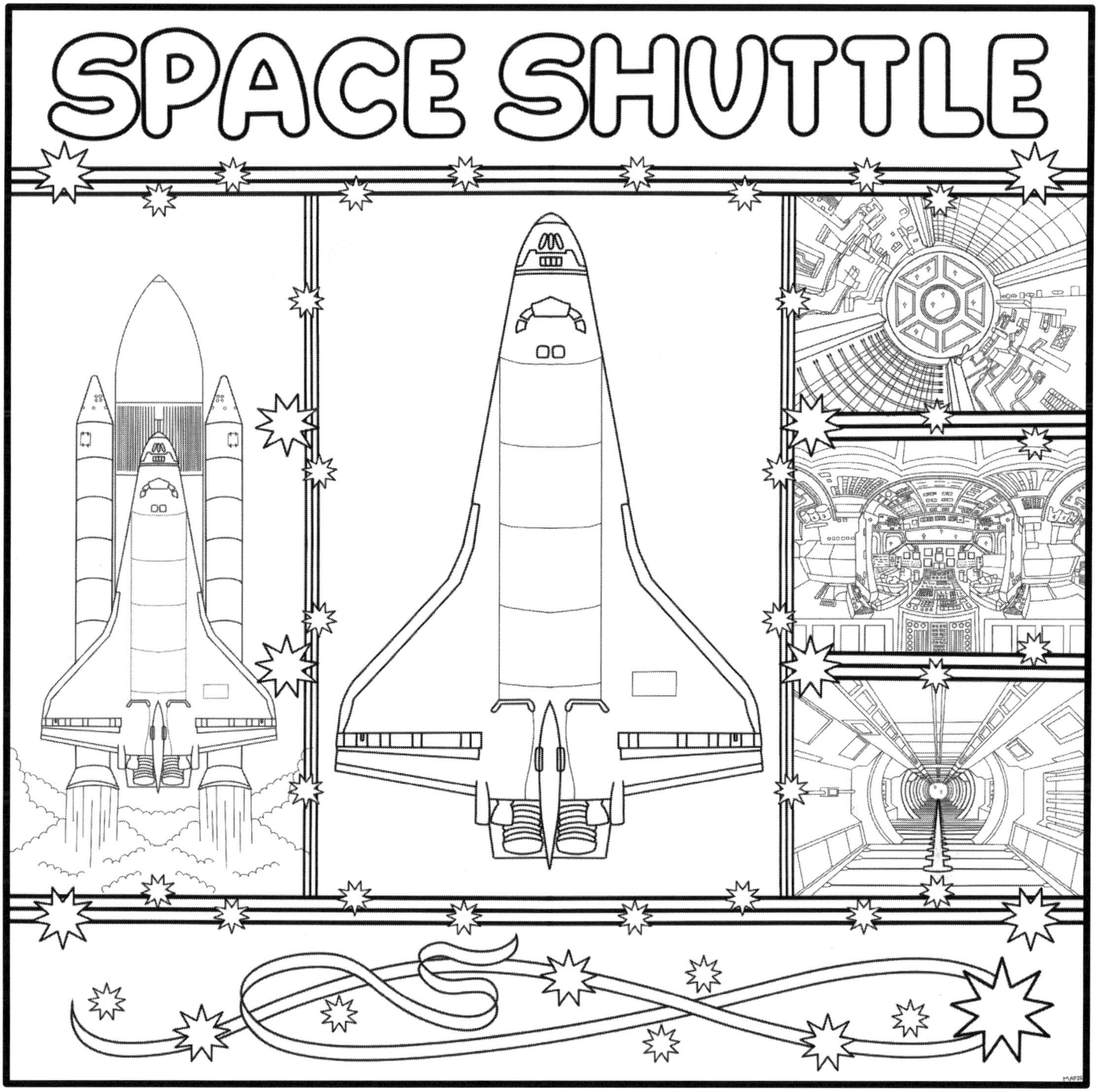

What made the space shuttle different from earlier spacecraft designs?	How did the space shuttle help scientists and engineers work in space?
Why was the development of a reusable spacecraft important?	What lessons did the space shuttle program provide for future space missions?

SPACE ROCKET

A space rocket is a powerful vehicle designed to launch spacecraft, satellites, and astronauts from Earth into space. Rockets work by producing thrust through the rapid expulsion of hot gases created by burning rocket fuel. This thrust pushes the rocket upward, allowing it to overcome Earth's gravity. Because there is no air in space, rockets do not rely on wings or air pressure and can function in the vacuum beyond Earth's atmosphere.

Space rockets are important because they are the only technology capable of carrying objects and people into space. They provide the force needed to escape Earth's gravitational pull and place spacecraft into orbit or send them toward other planets. Rockets can launch satellites for communication, weather monitoring, and navigation, as well as carry space probes, landers, and crewed spacecraft on exploration missions. Without rockets, modern space exploration and satellite technology would not be possible.

The history of rockets began long before space travel. Early rockets were developed for military and experimental purposes. In the 20th century, scientists such as Robert Goddard advanced rocket technology by developing liquid-fueled rockets capable of producing much greater thrust. During the mid-1900s, rocket technology advanced rapidly as nations competed during the space race. These developments led to the creation of rockets powerful enough to carry satellites and humans into space.

As space exploration progressed, rockets became more advanced and efficient. Multi-stage rockets were developed so that empty fuel tanks could be dropped during flight, reducing weight and increasing speed. These designs allowed spacecraft to reach orbit and travel deeper into space. Rockets were used to launch historic missions such as the first artificial satellites, human spaceflights, Moon landings, and space stations. Today, modern rockets continue to evolve with improved materials, guidance systems, and reusable components. Space rockets remain essential to exploration beyond Earth. They serve as the foundation of all space missions, enabling scientific discovery, global communication, and the exploration of distant worlds. By providing the power needed to reach space, rockets have transformed humanity's ability to study the universe and our place within it.

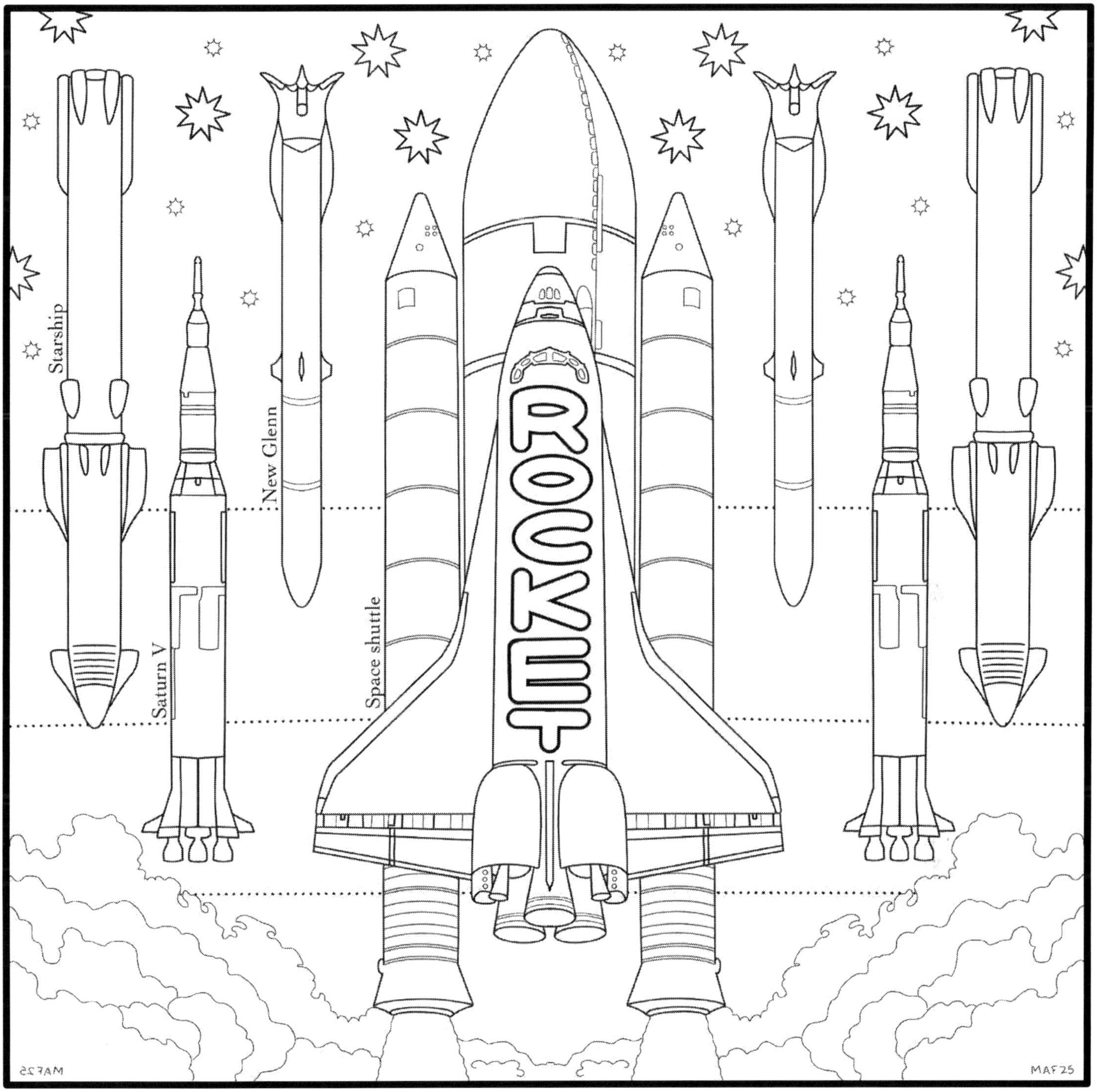

How does a space rocket produce the force needed to leave Earth?	Why are rockets essential for space exploration and satellite launches?
Describe one way rocket technology improved over time.	What are the advantages of using a multi-stage rocket?

ORRERY

An orrery is a mechanical or digital model that represents the positions and motions of celestial bodies, most commonly the Sun, planets, and their moons. In astronomy and science, an orrery is used to demonstrate how objects move in space, especially how planets orbit the Sun. Orreries are not built to scale in size or distance, but they accurately show relative motion, orbital paths, and timing.

The purpose of an orrery is to help scientists, students, and observers visualize complex astronomical motions that are difficult to understand from equations or diagrams alone. By watching an orrery in motion, it becomes easier to understand concepts such as orbital periods, planetary alignment, retrograde motion, eclipses, and the relationship between Earth's movement and seasons. Orreries turn abstract space concepts into observable mechanical motion.

The history of the orrery dates back to the early 18th century. These models were developed as astronomy advanced beyond naked-eye observation and mathematical prediction. Early orreries were hand-crafted mechanical devices driven by gears and cranks. They were often used in universities, observatories, and royal courts to explain the heliocentric model of the solar system, which places the Sun at the center rather than Earth. The name "orrery" comes from the Earl of Orrery, who owned one of the early models.

As scientific understanding improved, orreries became more accurate in representing orbital timing and planetary relationships. Advances in engineering allowed gears to be designed so that each planet completed an orbit in the correct relative time. In modern times, digital orreries and computer simulations have expanded these models far beyond physical limitations. Today's orreries can include thousands of objects, real-time data, elliptical orbits, and gravitational interactions, allowing scientists and students to model both our solar system and distant planetary systems.

Orreries remain important educational and scientific tools because they help bridge the gap between observation and theory. While telescopes allow astronomers to observe space directly, orreries help explain how celestial motions work overtime. By modeling orbital mechanics in a clear and interactive way, orreries continue to support the study of astronomy and deepen understanding of how objects move through space.

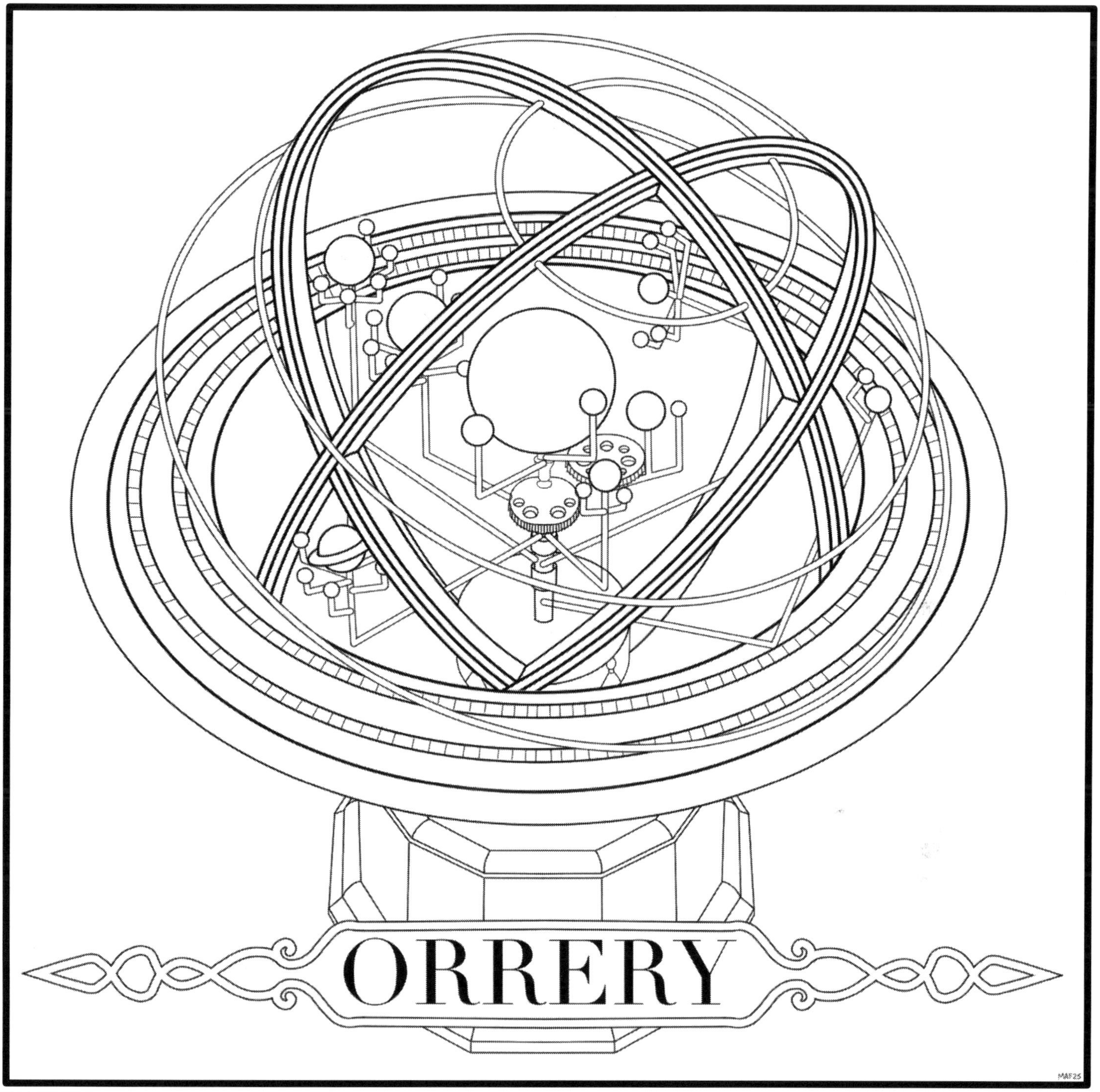

What is an orrery, and what does it represent in astronomy?	Why are orreries useful for understanding planetary motion and orbits?
How did early mechanical orreries help support the heliocentric model of the solar system?	How have modern advances changed the way orreries are used today?

UNIVERSE

The universe is everything that exists, including all space, time, matter, and energy. It contains all galaxies, stars, planets, moons, asteroids, radiation, and the vast regions of space between them. In astronomy, the universe is studied as a whole system governed by physical laws such as gravity, motion, and the behavior of energy. Scientists believe the universe began approximately 13.8 billion years ago with an event known as the Big Bang, when space itself began expanding from an extremely hot and dense state.

As the universe expanded and cooled, matter began to form structures. Gas and dust collected under the force of gravity to create stars and galaxies. Galaxies grouped together into clusters and superclusters, forming a large-scale structure sometimes described as a cosmic web. Within galaxies, stars formed planetary systems, including our own solar system. The universe is still expanding today, and distant galaxies continue to move farther apart over time.

The universe is not static but constantly changing. Stars are born, evolve, and die, sometimes in powerful explosions called supernovae that spread heavy elements into space. These elements become part of new stars, planets, and eventually living organisms. Black holes, neutron stars, and other extreme objects shape their surroundings through intense gravity and energy. Studying these processes helps astronomers understand how matter and energy interact under the most extreme conditions known.

Astronomers study the universe using telescopes and spacecraft that observe different forms of electromagnetic radiation, including visible light, infrared, X-rays, and radio waves. By measuring distances, motion, and light from distant objects, scientists can determine the age, size, and composition of the universe. Observations of the universe also provide clues about dark matter and dark energy, which make up most of the universe's total mass and energy but are not yet fully understood.

Through the study of space, we have explored planets, moons, stars, galaxies, orbits, light-years, and the tools that allow us to observe them. Each discovery reveals how connected the universe truly is, from the smallest particles to the largest cosmic structures. As space exploration and technology continue to advance, future missions will explore deeper into space, uncover new worlds, and answer questions we have not yet imagined. The universe invites curiosity, discovery, and exploration, reminding us that learning about space is also a way of understanding our place within it.

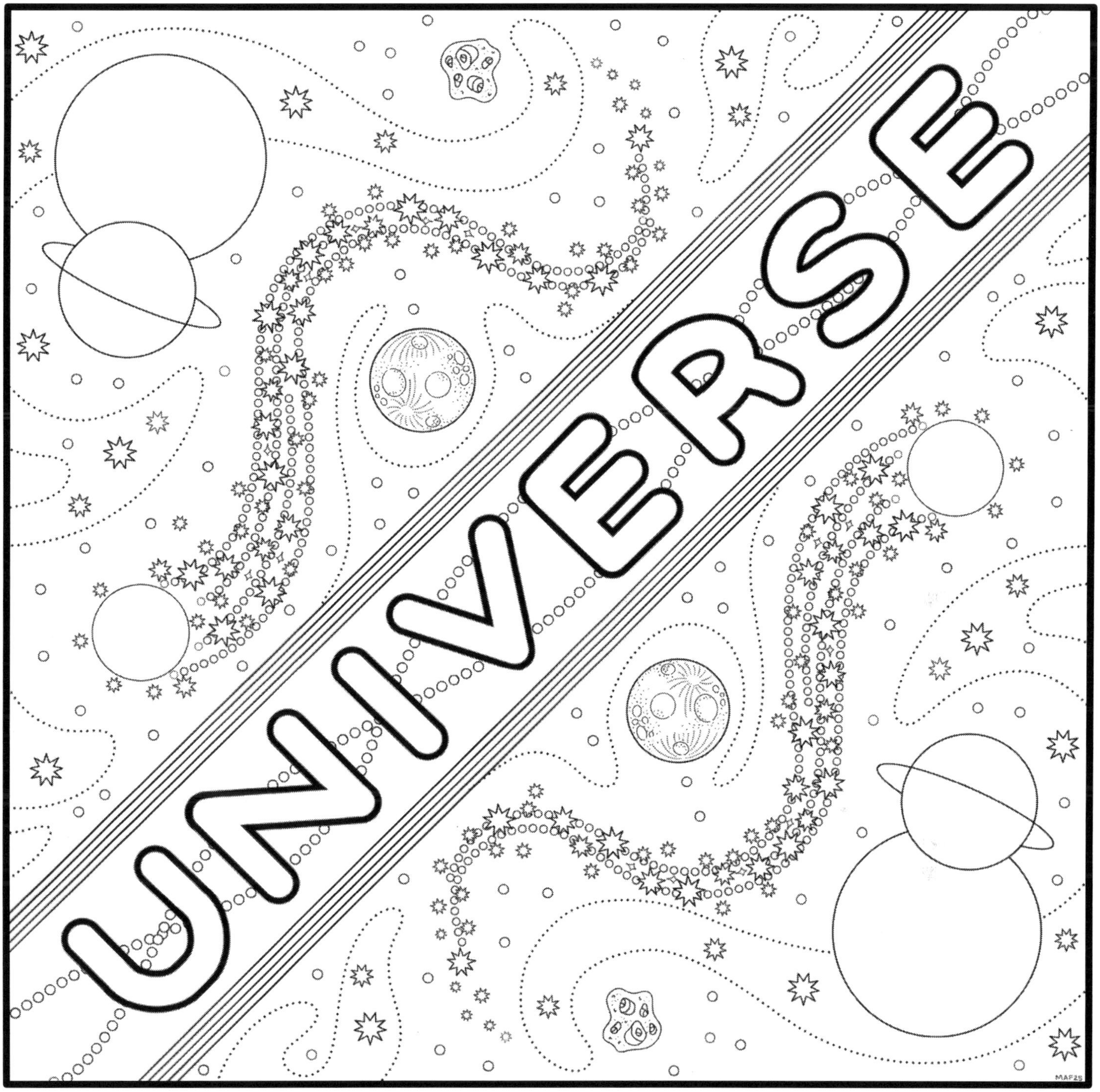

What is the universe, and what does it include according to astronomy?	How did structures such as stars and galaxies form after the beginning of the universe?
Why do astronomers say the universe is constantly changing?	What do scientists know about dark matter and dark energy?

Made in the USA
Coppell, TX
05 November 2026

11314461R00059